HELLENISTIC INFANTRY REFORM IN THE 160`S BC.

STUDIES ON THE HISTORY OF ANCIENT AND MEDIEVAL ART OF WARFARE

VOLUME V

NICHOLAS SEKUNDA

HELLENISTIC INFANTRY REFORM IN THE 160`S BC.

ŁÓDŹ 2001

Cover designed by
Emilia Wtorkiewicz-Marosik

Oficyna Naukowa MS
P.O. Box 126
90-965 Łódź 36
Poland

For details of all "Oficyna Naukowa MS" publications and titles in preparation please write to the above address. Volumes are distributed directly by the publisher. Prices are inclusive of postage by surface mail anywhere in the world.

ISBN 83-85874-04-6

Druk i oprawa:
Drukarnia Wydawnictw Naukowych S.A.
Łódź, ul. Żwirki 2

NICHOLAS SEKUNDA

HELLENISTIC INFANTRY REFORM IN THE 160`S BC.

OFICYNA NAUKOWA MS
ŁODŹ 2001

PREFACE

A special debt of thanks must go to Mariusz Mielczarek for his encouragement and help in the writing of this book. The completion of the manuscript would have been impossible without access to the Ashmolean Library and the John Rylands University Library Manchester, to both of which institutions I owe a great debt of thanks. Also many thanks to Richard Catling, Richard Ashton and Roger Brock who have helped with pursuing references on my behalf, and to Richard Brzezinski who has helped to save some of the English of the text.

No single academic convention on how to render Greek personal and location names in English has been established, and I have not adopted any firm policy in this work. In general I have tried to retain the Latinized English form for familiar names of known historical personalities such as Alexander or Pyrrhus, but I have tended to render the names of less familiar persons, for example individuals only appearing in papyri, in phonetic transcription from Greek to English. I trust that I have not rendered the name of any individuals in both systems in different parts of the text, but this is all too possible in the current confused situation. All dates are BC unless otherwise stated.

CONTENTS

ABBREVIATIONS

(a) Standard abbreviations

AAA	*'Αρχαιολογικὰ 'Ανάλεκτα ἐξ 'Αθηνῶν / Athens Annals of Archaeology.*
AHB	*Ancient History Bulletin.*
AJA	*American Journal of Archaeology.*
Arch. Delt.	*'Αρχαιολογικὸν Δελτίον.*
BCH	*Bulletin de correspondance hellénique.*
BGU	edd. U. Wilcken, W. Schubart, et al., *Aegyptische Urkunden aus den Staatlichen Museen zu Berlin, Griechische Urkunden* 1- (Berlin 1895-).
BSA	*The Annual of the British School at Athens.*
BSAAlex	*Bulletin de la Société (royale) d'Archéologie d'Alexandrie.*
BSOAS	*Bulletin of the School of Oriental and African Studies.*
Chiron	*Chiron.*
CP	*Classical Philology.*
CQ	*Classical Quarterly.*
CR	*Classical Review.*
FGrH	F. Jacoby, *Fragmente Griechischer Historiker* (Berlin, Leiden 1923-58).
FHG	C. Müller, *Fragmenta historicorum Graecorum* (Paris 1853-70).
GRBS	*Greek, Roman and Byzantine Studies.*
Hesp.	*Hesperia.*
ICret.	*Inscriptiones Creticae,* ed. M. Guarducci (Rome 1935-50).
IG	*Inscriptiones Graecae* (Berlin 1873-).
IGR	*Inscriptiones Graecae ad res Romanas pertinentes,* edd. R. Cagnat, G. Lafaye et al. (Paris 1906-11).
JEA	*Journal of Egyptian Archaeology.*
JRS	*Journal of Roman Studies.*

Kret. Khron.	*Kretika Khronika*
LGPN	*Lexicon of Greek Personal Names,* edd. P.M. Fraser, E. Matthews et al. (Oxford 1987-).
OCD²	*Oxford Classical Dictionary,* edd. N.G.L. Hammond, H.H. Scull ard (second edition, Oxford 1970).
OGIS	*Orientis Graeci inscriptiones selectae,* ed. W. Dittenberger (LeiP-zig 1903-5).
Opus. Athen.	*Opuscula Atheniensia.*
PBad.	*Veröffentlichungen aus den badischen Papyrus-Sammlungen,* edd. W. Spiegelberg, F. Bilabel, G. A. Gerhard 1-6 (Heidelberg 1923-38).
PFayum	Bernard P. Grenfell, Arthur S. Hunt, David G. Hogarth, *Fayûm towns & their Papyri* (London 1900).
P. Grenf.	B.P. Grenfell, *An Alexandrian erotic fragment and other Greek papyri chiefly Ptolemaic* (1896).
PP	W. Peremans, E. Van't Dack, *Prosopographia Ptolemaica* I-IX (Louvain 1950-1981).
PPetrie	*The Flinders Petrie Papyri,* edd. J. P. Mahaffy, J. G. Smyly 1-3 (Dublin 1891-1905).
PTebt.	*The Tebtunis Papyri,* edd. B.P. Grenfell, A.S. Hunt, J.G. Smyly et al. 1-3 (London 1902-1938).
PW	*Real-Encyclopädie der classischen Altertumswissenschaft,* edd. A.F. Pauly, G. Wissowa, W. Kroll et al. (Stuttgart 1894-1980).
Rev. arch.	*Revue archéologique.*
RÉG	*Revue des études grecques.*
Rev. Phil.	*Revue de Philologie.*
Riv. Fil.	*Rivista di filologia.*
SA	*Sovietskaya Arkheologia*
SB	*Sammelbuch griechischer Urkunden aus Ägypten,* edd. F. Preisigke, F. Bilabel et al. 1- (Strassburg, Berlin, Leipzig etc. 1915-).
SEG	*Supplementum Epigraphicum Graecum* 1- (1923-).
TAPA	*Transactions of the American Philological Association.*
UPZ	*Urkunden der Ptolemäerzeit,* ed. U. Wilcken, 1-2 (Berlin 1927, 1957).
ZPE	*Zeitschrift für Papyrologie und Epigraphik,*

(b) Non-Standard abbreviations. The following works listed in the Bibliography appear in the text in the following abbreviated forms:

Van't Dack, *Ptolemaica Selecta*

E. Van't Dack, *Ptolemaica Selecta. Études sur l'armée et l'administration lagides* (*Studia Hellenistica 29,* Louvain 1988).

Bar-Kochva, *Seleucid Army*

Bezalel Bar-Kochva, *The Seleucid Army. Organization and Tactics in the Great Campaigns* (Cambridge 1976).

Bar-Kochva, *Judas Maccabaeu*

Bezalel Bar-Kochva, *Judas Maccabaeus. The Jewish Struggle against the Seleucids* (Cambridge 1989).

Launey

Marcel Launey, *Recherches sur les Armées hellénistiques* I-II (Paris 1949, 1950).

Lesquier

Jean Lesquier, *Les Institutions militaires de l'Égypte sous les Lagides* (Paris 1911).

Walbank, *Commentary*

F. W. Walbank, *A Historical Commentary on Polybius* I (Oxford 1956), II (Oxford 1967), III (Oxford 1979).

LIST OF FIGURES IN TEXT

INTRODUCTION

The thesis advanced in this book is that both the Ptolemaic and Seleucid heavy infantry were re-armed and re-organized along Roman lines during the 160's BC. This was the first time that the military successes of Rome forced the Hellenistic world to accept the fact that the Hellenistic phalanx was no match to the Roman manipular tactical system, and that it was necessary to change the equipment, organization and tactics of their heavy infantry.

There is some evidence for previous Roman influence on the heavy infantry of the Hellenistic monarchies, particularly upon the army of Pyrrhus. This evidence will be discussed in Appendix 1. It is clear that this influence was sporadic and superficial and did not lead to a thorough reform of the Hellenistic phalanx.

The evidence for military reform comes from an extremely varied range of sources, literary, epigraphic, papyrological and representational. As the evidence is more conclusive for the Ptolemaic army, this will be dealt with first in Chapter 1. It is necessary to examine the evidence for the prior organization of the Ptolemaic army, then the epigraphic and papyrological evidence for reform will be discussed first as this type of evidence can give concrete dates for the change. Finally the various individual examples of representations, or possible representations, of Ptolemaic heavy infantry equipped in the Roman manner will be discussed. The discussion of the Ptolemaic evidence will conclude with a discussion of the Kasr el Harit shield, a shield of Roman type which was probably used by a Ptolemaic infantryman.

In Chapter 2 the evidence for reform of the Seleucid heavy infantry is discussed. The Seleucid evidence is much more sparse and difficult to interpret than the Ptolemaic, which is true of virtually all aspects of the Seleucid state. Nevertheless I believe it is possible to demonstrate that the Seleucid heavy infantry too was reformed along Roman lines in the 160s. The principal piece of evidence we possess is a literary reference to 5,000 infantrymen 'armed in the Roman manner' participating in the Daphne Parade in 166 BC.

The evidence for military reform along Roman lines in the Hellenistic armies has not previously been discussed. In 1911 Lesquier suggested that the title of the regiment in the Ptolemaic army changed from *syntagma* to *hêgemonia* at some point during the second century BC,[1] while in an article which originally appeared in 1963 Van't Dack noted that the rank of *taxiarchos* is only attested for the third century, and the rank of *hekatontarchos* is only attested in the second and first.[2] Both of these writers failed to connect such changes in titles of officers with a thorough reform of the Ptolemaic heavy infantry along Roman lines.

Some of the ideas advanced in this book have previously appeared in two titles of a popular nature, directed towards the general reader. These were published in Great Britain by Montvert Publications, namely *Seleucid and Ptolemaic Reformed Armies 168-145 BC. Volume 1: The Seleucid Army under Antiochus IV Epiphanes* (1994) and *Seleucid and Ptolemaic Reformed Armies 168-145 BC. Volume 2: The Ptolemaic Army under Ptolemy VI Philometor* (1995). Lack of space, especially in the first of these two volumes, precluded the inclusion of all the material initially intended. Similarly the necessity of making these books as accessible as possible to the general public did not allow the inclusion of the full academic apparatus. Needless to say, research work carried out since that time has also added to the information available at that time, and my ideas of what happened in the infantry reforms of the 160s have changed somewhat.

---

[1] Lesquier p. 96.
[2] Van't Dack, *Ptolemaica Selecta* p. 59.

CHAPTER 1

MILITARY REFORM IN THE PTOLEMAIC ARMY

Section 1 a

The Structure of the Early Ptolemaic Army

Before examining the evidence for new structures and ranks introduced into the Ptolemaic infantry in the 160s, we shall first outline the structure of the army prior to the reform. The Ptolemaic army was essentially divided into two components, the standing army and the cleruch army. The standing army comprised the regiments of guards which protected the person of the king and the court, and the regiments of mercenaries stationed in garrisons throughout the Empire. The cleruchic army was a 'territorial' army. From the reign of Ptolemy I Soter onwards, and especially from the reign of Ptolemy II Philadelphos, ex-soldiers were allotted plots (*klêroi* – hence the appellation *klêrouchoi* given to the settlers), in return for the liability to perform military service in time of war, if called upon.[1] An individual cleruch held a particular rank, and belonged to a particular regiment in this territorial army, and upon mobilization a fully formed army should, in theory, have constituted itself. At first, for example during the Third Syrian War, the system seems to have worked reasonably well, but by the end of the third century it was already beginning to malfunction.

As well as their rank and regiment, individual cleruchs have all sorts of other titles in the papyri, which are, as yet, not understood with any certainty. It is difficult

[1] On the cleruchic system see D. J. Crawford, *Kerkeosiris. An Egyptian Village in the Ptolemaic Period* (Cambridge 1971) pp. 55-85. For a complete bibliography see E. Van't Dack in *Armées et fiscalité dans le Monde antique, Colloques Nationaux du C.N.R.S No, 936* (Paris 1977) p. 80-1.

to be certain how much military practices in the standing army and in the cleruchic army would have been identical. It is probable, however, that the two forces were identical in their regimental structures and ranks. From papyrological material which is, in the main, relevant to the cleruchic army, a rough outline can be given for the organization of the Ptolemaic army during the third century. The military ranks held by individuals appearing in the Ptolemaic papyri have been collated in the work *Prosopographia Ptolemaica.* In the text below the numbers have been given as they appear in this work in brackets after each rank.

As with all the early Hellenistic armies, the Ptolemaic army largely followed the organizational structures of the Macedonian army as they stood at the end of the reign of Alexander the Great.[2] One factor which has hampered research into Hellenistic armies has been the prevailing concept that the organization of Alexander's army was unchanged, that it was copied without substantial variation by all subsequent Hellenistic armies, and that the outline of an ideal army given in the Greek *Taktika* of Asclepiodotus, Aelian and Arrian reflects the organization of the Macedonian army under Alexander. In fact one can detect organizational reforms in the Macedonian army throughout Alexander's reign. While all Hellenistic armies were originally formed from regiments surviving the break-up of Alexander's army, and so all Hellenistic armies originally inherited the same organizational structures, they all evolved over time. Finally it is clear that the Greek *Taktika*, the earliest of which was written in the first century BC, borrow their terminology from a number of sources, but these sources are late Hellenistic not Alexandrine (see Appendix B).

The cavalry under Alexander was organized into squadrons, or *ilai*, which were then grouped into brigades, or *hipparchiai*. Alexander crossed over into Asia with 1,800 Macedonian Companion cavalry (Diod. 17. 17. 4). We know that there were eight squadrons of Companion cavalry. It seems probable that the so-called 'Royal' squadron of the Companions was double strength, and that the normal strength of a squadron would have been about 200 men.[3] The Ptolemaic cavalry was similarly divided into numbered hipparchies, and then into *ilai*. Hipparchies numbered from one to ten are currently attested, with the exception of hipparchies numbers six and nine which are not mentioned in any documents yet published.[4] In the winter of 331/0 Alexander divided each of the cavalry *ilai* into two *lochoi* (Arr., *Anab.* 3. 16. 11). A Ptolemaic cavalry *lochos* may also have existed, and a cavalry *dekania* commanded by a *dekanikos*.[5] The cavalry was, it seems, largely unaffected by the military reforms to be discussed below, for the cavalry continued to be organized into hipparchies and *ilai*. The rank of *dekanikos* is not however

[2] Thus Lesquier p. v.

[3] N. G. L. Hammond and G. T. Griffith, *A History of Macedonia. Volume II 550-336 B.C.* (Oxford 1979) p. 411.

[4] Van't Dack, *Ptolemaica Selecta* p. 53.

[5] Lesquier pp. 90-1.

attested after 165.

Curtius (5. 2. 3) tells us that in the winter of 331/0 Alexander reorganized the infantry into *chiliarchiai* of 1,000 and that previously there had only been *lochoi* of 500 men.[6] *Lochoi* are later recorded as a sub-division of the infantry chiliarchy (4. 2. 1). The ranks of *chiliarchoi* 'commanders of a thousand', and *pentakosiarchoi* 'commanders of five hundred' are both attested in the later army of Alexander (Arr., *Anab.* 7. 25. 6; Plut., *Vit Alex.* 76. 6). They were presumably in command of infantry units with establishment strengths, based on a file of 16 men, of 1,024 and 512 men respectively. The third-century Ptolemaic papyri likewise mention officers who hold the ranks of *chiliarchos* (*PP* 2290-2301) and *pentakosiarchos* (*PP* 2302-2320), and so we may safely conclude that the regimental structure of Alexander's army continued in the Ptolemaic army throughout the third century. The Foot-Companions of Alexander's army had originally been divided into *taxeis* each of three *lochoi*.[7] In the Ptolemaic army we also hear of *taxiarchoi* 'commanders of companies' (*PP* 2288-9), who may have commanded an infantry sub-unit or unit on a different level in the organizational hierarchy.

Whenever the rank of *dekanikos*, or 'commander of ten', appears in the papyri (*PP* 2273-2287) and the unit is given, the individual bearing the rank of *dekanikos* is frequently found to be serving in the cavalry. As well as the *dekanikoi* we also hear of *dimoiritai* 'double-pay men' in the papyri.[8] Both these two ranks closely parallel those in the infantry file of sixteen men in the Macedonian army under Alexander. The organization of the file under Alexander is given in Arrian, *Anab.* 7. 23. 3-4.[9] For traditional reasons the file was called a *dekas*, or 'ten', even though it numbered sixteen men.[10] Presumably the Macedonian file had once numbered ten men in the distant past, but, when the number of the file was expanded to sixteen men in line with current Greek military practice, the old term *dekas* was retained. The commander of the *dekas* was called a *dekadarchos*, and the *dekas* also had a *dimoiritês* and two *dekastatêroi* or 'ten-stater men'. This information is given in a discussion of how in the last year of Alexander's reign the *dekas* was reorganized to accommodate native Persians. Before this reform,

---

[6] Cf. Augustin Spendel, *Untersuchungen zum Heerwesen der Diadochen* (Breslau/Wroclaw 1915) p. 35 who notes that the chiliarchy is previously attested in the corps of hypaspists, which preceded the wholesale restructuring of the infantry.

[7] See Robert D. Milns in ed. E. Badian, *Alexandre le Grand. Image et réalité. Entretiens Hardt XXII* (Geneva 1976) p. 102; N. G. L. Hammond and G. T. Griffith, *A History of Macedonia. Volume II 550-336 B.C.* (Oxford 1979) pp. 416 n. 2, 419.

[8] *PP* 3878, 3921 (dated to 203/2 BC) and 3994.

[9] For a detailed discussion of this and other passages giving the strength of the phalanx file as 16 see Alfred v. Domeszewski, *Die Phalangen Alexanders und Caesars Legionen* (= *Sitzungsberichte der Heidelberger Akademie der Wissenschaften, Philosophisch-historische Klasse 1925/6. 1*, Heidelberg 1926) p. 24-8.

[10] N. G. L. Hammond and G. T. Griffith, *A History of Macedonia. Volume II 550-336 B.C.* (Oxford 1979) p. 419-20.

however, it is reasonable to suppose that the *dekadarchos* and the *dimoiritês* stood at the front of the two half-files of eight men, and the two *dekastatêroi* stood at the back of either half-file. While the *dekastatêroi* are unattested in the papyri the *dimoiritai* are. The term *dekanikos* might be simply an alternative title for the *dekadarchos*. It seems reasonable to conclude that the Ptolemaic infantry file was called a *dekania* and that *dekanikos* was also an infantry rank.[11]

All but one references to *dekanikoi*, whether infantry or cavalry, date to the third century. This implies that an organizational reform had been carried out in the smallest sub-units of the cavalry and infantry alike. Similarly the ranks of *chiliarchos* and *pentakosiarchos* are only found in papyri of the third century and not of the second. This implies that the infantry was thoroughly reformed. The last reference to any soldier holding the rank of *dekanikos* comes in a manumission document from Krokodilopolis in the Arsinoite Nome dating to 19 Apellaios/Payni 165 BC.[12] In it one Adrastos son of Adrastos, a native of the Euboean city of Hestiaia, and so presumably a mercenary soldier, *dekanikos* of the 'Regiment formerly commanded by Noumenios' ([τ]ῶ[ν πρότερον] Νουμηνίου) makes a declaration under oath.[13] The commander who gave his name to the regiment, and who is called its eponymous commander in the conventions of modern writing, has been identified with Noumenios son of Herakleiodoros of Alexandria. He is known to have served earlier as *stratêgos* of the Thebaïd, but was summoned to Alexandria late in 170, probably in preparation for the war. Noumenios left military service late in 168 to preside over the embassy to Rome, and he eventually rose to the rank of *epistolographos*, or head of chancellery under Ptolemy VIII Euergetes II.[14] It is extremely interesting to note that the regiment formerly commanded by Noumenios remained without an officially appointed successor to the eponymous command for at least two years.

A date of July 165 BC therefore constitutes a possible *terminus post quem* for reform in the Ptolemaic infantry. The date is not absolutely certain, however, for the unit commanded by Noumenios may have been a cavalry unit, and it is theoretically possible that the rank disappeared in the cavalry later than it did in the infantry. In the opinion of the present author, however, this seems unlikely. We shall now turn to examine the evidence, epigraphic and papyrological, for a new structure appearing in the Ptolemaic infantry. All the evidence which can be firmly dated has a date later than 165 BC.

---

[11] Cf. Lesquier p. 92.

[12] The month of Apellaios/Payni is the tenth month of the Egyptian year, which began around 1 October in the year 166/5, thus the month Apellaios/Payni would equate to July 165 (Alan E. Samuel, *Greek and Roman Chronology. Calendars and Years in Classical Antiquity* (= edd. Iwan von Müller, Walter Otto, Herman Bengtson, *Handbuch der Altertumswissenschaft I, 7;* Munich 1972) p. 145.

[13] *PTebt.* III p. 267-8 no. 811 line 13-14; *PP* 2273.

[14] W. Peremans & E. Van't Dack, *Prosopographica* (Louvain 1953) p. 50 n. 6; Walbank, *Commentary* III pp. 439, 453.

Section 1 b

The 'Lefebvre' Stele from Hermopolis

Probably the most important single piece of evidence, and therefore the one with which we will start our discussion, is the earliest of three inscriptions listing military units coming from Ashmunein, the ancient city of Hermopolis Magna, which were deposited by the local garrison. The inscription is either known as the Lefebvre Stele after its first publisher Gustave Lefebvre or as 'Hermopolis Stele L' (Figure 1).[15] The inscription, broken at both top and bottom, lists at least some of the military contingents of the Hermopolis garrison. Thus it is of singular importance in giving us the organization of the units listed.

Our understanding of the inscription is hampered firstly by the fact that the stele (and the lists of names it contains) is broken at the top and at the bottom. Because the inscription is broken off at the top, we do not know what its precise purpose was, but it is most probably dedicatory in nature.[16] Furthermore, while we can assume that the Lefebvre Stele records some sort of dedication, we do not know its precise nature and therefore we cannot be absolutely certain that when a sub-unit is listed it records all the members of that sub-unit present in the garrison, as opposed to, for example, only those who had made a contribution to the temple. The former explanation nevertheless seems the most likely.

The date of the inscription has not yet been established with complete certainty. It may be possible to identify the officer Pasines who, along with another individual called Dryton, is mentioned as commanding a unit in line III, 19.[17] As these officers give their name to the unit they command they are called eponymous officers. Pasines is also named in a papyrus, which probably also comes from Hermopolis Magna, and which mentions one Korax, son of Dionysios, a standard-bearer of those under Pasines. The papyrus has been dated to 143/2 BC, and so we would expect the Lefebvre Stele to be close in date. There is a reference in the stele in line I, 60 to persons who have received an amnesty, or *philanthropa*. This amnesty will probably be the amnesty made by Ptolemy VIII Euergetes II in 145/4 BC,[18]

---

[15] Gustave Lefebvre, 'Inscription Grecque d'Ashmounein' *BSAAlex* 10 (1908) pp. 187-195; E. Breccia, *Catalogue général des antiquités égyptiennes du Musée d'Alexandrie. Iscrizioni greche e latine* (Cairo 1911) no. 44a; *SB* 599. For bibliography of discussion of this inscription see Étienne Bernand, *Inscriptions grecques d'Hermoupolis Magna et de se nécropole* (Cairo 1999) no. 4 pp. 21-35 (bibliography p. 21).

[16] Jan Krzysztof Winnicki, *Ptolemäerarmee in Thebais* (Wroclaw 1978) p. 12.

[17] *PP* 1972-3.

[18] Marie-Thérèse Lenger, *Corpus des Ordonnances des Ptolémées* (Brussels 1964) pp. 41-3.

Figure 1. The 'Lefebvre Stele' from Hermopolis (photograph: P. M. Fraser).

1re COLONNE.
(vacat)
?
[illegible]ΗΜΗΤΡ[illegible]Σ
[illegible]ΤΟΛΕΜΑΙΟΥ ΑΙΑ[illegible]
ΑΓΟΛΛΩΝΙΟΣ ΔΩ[illegible]ΟΥ
[illegible]ΟΜΑΝΟΥΚΑΙΤΩΝΑΛΛΩΝ
ΚΗΡΥΞ
ΜΗΝΟΔΩΡΟΣ ΜΗΝΟΔΩΡΟΥ
ΣΗΜΕΙΟΦΟΡΟΣ
ΘΕΟΔΩΡΟΣ ΗΡΑΚΛΕΙΤΟΥ
ΕΚΑΤΟΝΤΑΡΧΟΣ
ΔΗΜΗΤΡΙΟΣ ΑΓΟΛΛΩΝΙΟΥ
ΤΗΡΗΣ ΤΗΡΟΥ
ΑΓΟΛΛΩΝΙΟΣ ΑΓΟΛΛΩΝΙΟΥ
ΗΛΙΟΔΩΡΟΣ ΓΤΟΛΕΜΑΙΟΥ
ΕΡΜΟΚΡΑΤΗΣ ΣΩΣΤΡΑΤΟΥ
ΔΙΟΝΥΣΙΟΣ ΦΙΛΟΚΡΑΤΟΥ
ΑΓΑΘΟΚΛΗΣ ΑΓΑΘΟΚΛΕΟΥΣ
ΔΙΟΣΚΟΥΡΙΔΗΣ ΔΙΟΣΚΟΥΡΙΔΟΥ
ΓΤΟΛΕΜΑΙΟΣ ΑΝΔΡΟΝΙΚΟΥ
ΑΤΘΑΝΙΑΦΑΝΤΟΣΚΑΙΣΘΕΝΕΛΑΟΥ
ΘΕΩΝ ΔΗΜΗΤΡΙΟΥ
ΒΙΛΟΣ ΒΙΛΟΥ
ΚΟΜΑΝΟΥΚΑΙΤΩΝΑΛΛΩΝ
ΑΓΟΛΛΟΔΟΤΟΣ ΑΓΟΛΛΟΔΟΤΟΥ
ΑΤΘΑΝΙΑΦΑΝΤΟΣΚΑΙΣΘΕΝΕΛΑΟΥ
ΣΗΜΕΙΟΦΟΡΟΣ
ΑΝΔΡΗΜΩΝ ΒΑΛΑΚΡΟΥ
ΓΕΝΤΗΚΟΝΤΑΡΧΟΣ
ΔΙΟΝΥΣΙΟΣ ΓΤΟΛΕΜΑΙΟΥ
ΑΘΗΝΟΔΩΡΟΣ ΑΘΗΝΟΔΩΡΟΥ
ΔΙΟΝΥΣΙΟΣ ΔΙΟΝΥΣΙΟΥ
ΑΓΟΛΛΩΝΙΟΣ ΑΓΟΛΛΩΝΙΟΥ
ΑΕΤΟΣ ΣΩΚΡΑΤΟΥ
ΑΙΣΧΡΙΩΝ ΓΤΟΛΕΜΑΙΟΥ
ΔΙΟΝΥΣΟΔΩΡΟΣ ΔΙΟΝΥΣΟΔΩΡΟΥ
ΚΑΙΤΩΝΓΡΟΤΕΡΟΝ
ΜΕΤΑΔΡΥΤΩΝΟΣΗΓΕΜΩΝ
ΕΓΑΝΔΡΩΝΚΑΙΦΡΟΥΡΑΡΧΟΣ
ΤΙΜΟΚΛΗΣ ΤΙΜΟΚΛΕΟΥΣ
ΕΞΩΤΑΞΕΩΝ ΗΓΕΜΟΝΕΣ
ΑΡΙΣΤΑΡΧΟΣ ΔΙΟΓΝΗΤΟΥ
ΚΡΙΤΩΝ ΑΡΧΑ
ΕΞΩΤΑΞΕΩΝ
[illegible]ΡΙΣΤΟΜΑΧΟΣ ΑΡΙΣΤΟΜΑΧΟΥ
[illegible]ΕΜΟΝΩΝΥΙΟΙ
[illegible]ΟΝΥΣΙΟΣ ΔΙΟΝΥΣΙΟΥ
[illegible]ΑΝΤΕΛΗΣ ΔΙΟΝΥΣΙΟΥ
ΔΙΟΓΕΝΗΣ ΔΙΟΝΥΣΙΟΥ
ΓΡΑΜΜΑΤΕΥΣ ΣΥΝΤΑΞΕΩΣ
ΕΥΔΟΞΟΣ ΤΙΜΟΚΛΕΟΥΣ
ΜΕΤΑΓΛΑΟΜΑΧΟΥ
ΑΡΤΕΜΙΔΩΡΟΣ ΑΝΤΙΓΑΤΡΟΥ
ΑΓΟΤΗΣΘΗΒΑΙΔΟΣ
ΔΗΜΗΤΡΙΟΣ ΑΝΤΙΓΑΤΡΟΥ
ΜΗΝΟΦΙΛΟΥ
ΗΡΑΚΛΕΙΔΗΣ ΛΕΩΝΙΔΟΥ
ΚΡΗΤΩΝΑΡΙΣΤΟΚΑΡΤΟΥΓΟΡΤΥΝΙΟΥ
ΒΑΛΑΚΡΟΣ ΔΙΟΝΥΣΙΟΥ
ΚΥΡΗΝΑΙΚΩΝ ΑΝΔΡΟΝΙΚΟΥ
ΑΓΟΛΛΟΦΑΝΗΣ ΙΑΣΟΝΟΣ
ΤΩΝΤΑΦΙΛΑΝΘΡΩΓΑΕΧΟΝΤΩΝ
ΑΡΧΥΓΗΡΕΤΗΣ ΞΕΝΙΚΟΥ
ΣΩΓΑΤΡΟΣ ΚΑΣΣΙΟΔΩΡΟΥ
ΑΣΚΛΗΓΙΑΔΗΣ ΣΤΡΑΤΩΝΟΣ
ΓΟΣΣΙΚΡΑΤΗΣ ΓΟΣΣΙΚΡΑΤΟΥ
ΔΑΜΟΚΡΑΤΗΣ ΤΙΜΟΚΡΑΤΟΥ
ΦΑΙΝΕΜΑΧΟΣ ΔΙΚΑΙΑΡΧΟΥ
ΑΓΟΛΛΩΝΙΟΣ ΔΙΟΝΥΣΙΟΥ
ΦΙΛΙΣΤΟΣ ΘΕΟΔΟΣΙΟΥ
ΓΟΛΙΤΙΚΩΝ
ΑΛΕΞΑΝΔΡΟΣ ΓΤΟΛΕΜΑΙΟΥ
ΕΙΤΟΥΡΓ[illegible]Σ ΜΝΑΣΕΟΥ
ΑΓΑΘΟ[illegible]ΛΗΣ ΕΛΛΑΝ

2e COLONNE.
(vacat)
[illegible]ΩΣΟΣ ΣΩΤΑΔΟΥ
ΑΓΟΛΛΩΝΙΟΣ ΜΕΝΟΙΤΙΟΥ
ΑΚΕΣΩΝ ΑΚΕΣΩΝΟΣ
ΦΙΛΑΜΜΩΝ ΔΙΟΝΥΣΙΟΥ
ΗΡΑΚΩΝ ΦΙΛΙΣΚΟΥ
ΔΙΟΣΚΟΥΡΙΔΗΣ ΔΙΟΣΚΟΥΡΙΔΟΥ
ΓΤΟΛΕΜΑΙΟΣ ΤΙΜΩΝΟΣ
ΚΥΔΙΓΓΟΣ ΔΙΟΣΚΟΥΡΙΔΟΥ
ΑΓΑΘΟΚΛΗΣ ΞΕΝΟΤΙΜΟΥ
ΙΞΑΝΘΙΓΓΟΣ ΙΑΣΟΝΟΣ
ΑΡΙΣΤΟΜΑΧΟΣ ΑΡΙΣΤΟΜΑΧΟΥ
ΓΡΑΞΙΑΣ ΑΙΣΧΡΙΩΝΟΣ
ΑΓΟΛΛΩΝΙΟΣ ΝΙΚΑΝΟΡΟΣ
ΓΕΝΤΗΚΟΝΤΑΡΧΟΣ
ΕΡΜΙΓΓΟΣ ΛΕΟΝΤΟΣ
ΓΕΤΑΛΟΣ ΤΙΜΟΚΡΑΤΟΥ
ΞΕΝΟΔΑΜΟΣ ΑΙΣΧΡΙΩΝΟΣ
ΓΤΟΛΕΜΑΙΟΣ ΔΙΟΝΥΣΙΟΥ
ΓΥΘΙΩΝ ΔΗΜΗΤΡΙΟΥ
ΓΕΝΤΗΚΟΝΤΑΡΧΟΣ
ΓΥΘΕΑΣΝ ΝΕΟΓΤΟΛΕΜΟΥ
ΑΓΟΛΛΩΝΙΟΣ ΑΡΓΑΙΟΥ
ΛΥΚΟΣ ΑΝΔΡΩΝΟΣ
ΑΣΚΛΗΓΙΟΔΩΡΟΣ ΓΥΡΡΙΟΥ
ΤΗΡΗΣ ΤΗΡΟΥΣ
ΔΙΟΓΝΗΤΟΣ ΚΛΕΩΝΟΣ
ΣΤΡΑΤΩΝ ΑΣΚΛΗΓΙΑΔΟΥ
ΘΕΩΝ ΑΓΟΛΛΩΝΙΟΥ
ΗΡΑΚΛΕΙΔΗΣ ΕΙΣΙΔΩΡΟΥ
ΔΗΜΗΤΡΙΟΣ ΑΜΜΩΝΙΟΥ
ΑΡΧΙΛΟΧΟΣ ΓΥΡΡΟΥ
ΦΙΛΩΝ ΣΥΜΜΑΧΟΥ
ΘΕΟΜΒΡΟΤΟΣ ΓΕΤΑΛΟΥ
ΓΤΟΛΕΜΑΙΟΣ ΦΙΛΑΜΜΟΝΟΣ
ΑΣΚΛΗΓΙΑΔΗΣ ΑΓΟΛΛΩΝΙΟΥ
ΑΝΤΙΜΑΧΟΣ ΣΩΓΥΡΟΥ
ΛΟΒΙΟΣ ΛΟΒΙΟΥ
ΑΓΟΛΛΩΝΙΟΣ ΓΡΩΤΟΓΕΝΟΥ
ΓΤΟΛΕΜΑΙΟΣ ΔΙΔΥΜΟΥ
ΝΟΥΙΟΣ ΝΟΥΙΟΥ
ΑΓΟΛΛΩΝΙΟΣ ΣΑΤΥΡΟΥ
ΚΟΘΩΝ ΚΟΘΩΝΟΣ
ΔΙΟΣΚΟΥΡΙΔΗΣ ΚΟΘΩΝΟΣ
ΗΡΑΚΛΕΩΝ ΑΓΟΛΛΩΝΙΟΥ
ΑΙΑΣ ΑΘΗΝΑΔΟΥ
ΓΟΛΕΜΩΝ ΣΤΡΑΤΩΝΟΣ
ΦΙΛΩΝ ΦΙΛΩΝΟΣ
ΗΡΑΚΛΕΙΤΟΣ ΑΡΤΕΜΙΔΩΡΟΥ
ΑΣΚΛΗΓΙΑΔΗΣ ΑΓΟΛΛΩΝΙΟΥ
ΚΑΛΛΩΝ ΚΑΛΛΩΝΟΣ
ΓΕΝΝΑΙΟΣ ΓΕΝΝΑΙΟΥ
ΑΝΤΙΜΑΧΟΣ ΕΡΜΙΓΓΟΥ
ΓΕΝΤΗΚΟΝΤΑΡΧΟΣ
ΛΕΩΝΙΔΗΣ ΑΓΟΛΛΩΝΙΟΥ
ΔΙΔΥΜΟΣ ΕΡΜΟΓΕΝΟΥ
ΦΙΛΟΞΕΝΟΣ ΓΤΟΛΕΜΑΙΟΥ
ΑΝΤΙΦΙΛΟΣ ΔΙΟΝΥΣΙΟΥ
ΗΡΩΙΔΗΣ ΗΡΩΙΔΟΥ
ΣΑΡΑΓΙΩΝ ΔΙΟΝΥΣΙΟΥ
ΜΕΝΑΝΔΡΟΣ ΜΕΝΑΝΔΡΟΥ
ΟΥΡΑΓΟΣ
ΓΕΡΙΓΕΝΗΣ ΑΛΕΞΙΤΙΜΟΥ
ΔΙΑΔΟΥΚΑΙ
ΤΩΝ ΥΙΩΝ
ΚΗΡΥΞ
ΔΙΟΦΑΝΤΟΣ ΙΑ[illegible]Σ
(vacat)

3e COLONNE.
(vacat)
ΕΓΙΜΑΧΟΣ ΓΥ[illegible]ΡΟΥ
ΜΕΓΑΛΟΚΛΗΣ ΜΕΓΑΛΟΚΡΑΤΟΥ
ΑΓΝΟΘΕΟΣ ΜΕΣΤΟΥ
ΜΕΛΩΝ ΚΛΕΟΜΑΧΟΥ
ΜΑΙΑΝΔΡΙΟΣ ΓΡΕΓΕΛΑΟΥ
ΑΓΟΛΛΩΝΙΟΣ ΑΡΙΣΤΟΦΑΝΟΥ
ΔΙΟΝΥΣΙΟΣ ΕΥΞΙΘΕΟΥ
ΓΕΙΣΙΣΤΡΑΤΟΣ ΓΤΟΛΕΜΑΙΟΥ
ΙΓΓΑΛΟΣ ΑΡΙΣΤΟΜΑΧΟΥ
ΓΕΝΤΗΚΟΝΤΑΡΧΟΣ
ΝΕΩΝ ΜΗΤΡΟΔΩΡΟΥ
ΕΓΑΙΝΕΤΟΣ ΘΟΑΝΤΟΣ
ΑΓΟΛΛΟΦΑΝΗΣ ΚΛΕΩΝΥΜΟΥ
ΣΩΓΕΝΗΣ ΕΜΜΕΝΙΔΟΥ
ΔΗΜΗΤΡΙΟΣ ΚΛΕΟΣΘΕΝΟΥ
ΑΡΧΙΑΣ ΚΛΕΟΞΕΝΟΥ
ΟΥΡΑΓΟΣ
ΑΓΟΛΛΟΔΩΡΟΣ ΑΡΙΣΤΟΜΑΧΟΥ
ΓΑΣΙΝΟΥΚΑΙΔΡΥΤΩΝΟΣ
ΚΗΡΥΞ
ΔΡΑΚΩΝ ΗΓΗΣΙΟΥ
ΣΗΜΕΙΟΦΟΡΟΣ
ΜΙΚΚΟΣ ΓΤΟΛΕΜΑΙΟΥ
ΕΚΑΤΟΝΤΑΡΧΟΣ
ΓΤΟΛΕΜΑΙΟΣ ΤΡΥΦΩΝΟΣ
ΓΕΝΤΗΚΟΝΤΑΡΧΟΣ
ΣΩΙΛΟΣ ΣΩΙΛΟΥ
ΙΑΣΩΝ ΙΑΣΟΝΟΣ
ΓΤΟΛΕΜΑΙΟΣ ΑΡΤΕΜΙΔΩΡΟΥ
ΔΙΔΥΜΟΣ ΔΗΜΗΤΡΙΟΥ
ΓΡΩΤΟΓΕΝΗΣ ΕΥΦΡΑΝΟΡΟΣ
ΕΡΜΙΑΣ ΘΕΩΝΟΣ
ΓΡΩΤΟΓΕΝΗΣ ΚΛΕΟΜΑΧΟΥ
ΙΓΓΑΛΟΣ ΦΑΝΟΚΡΑΤΟΥ
ΓΑΡΜΕΝΙΩΝ ΑΡΙΣΤΟΜΑΧΟΥ
ΑΓΟΛΛΩΝΙΟΣ ΑΓΟΛΛΩΝΙΟΥ
ΓΤΟΛΕΜΑΙΟΣ ΑΜΥΝΤΟΥ
ΑΓΟΛΛΩΝΙΟΣ ΑΓΟΛΛΩΝΙΟΥ
ΓΡΩΤΕΑΣ ΕΓΙΚΡΑΤΟΥ
ΦΙΛΟΔΗΜΟΣ ΚΑΛΛΙΚΡΑΤΟΥ
ΑΡΙΣΤΟΜΑΧΟΣ ΑΡΙΣΤΟΜΑΧΟΥ
ΚΑΛΛΙΣΤΡΑΤΟΣ ΚΑΛΛΙΣΤΡΑΤΟΥ
ΕΡΜΟΓΕΝΗΣ ΕΥΓΟΛΕΜΟΥ
ΕΡΜΟΦΙΛΟΣ ΗΡΑΚΛΕΟΔΩΡΟΥ
ΔΙΟΔΩΡΟΣ ΜΕΝΑΝΔΡΟΥ
ΕΡΜΟΦΙΛΟΣ ΚΕΦΑΛΩΝΟΣ
ΓΤΟΛΕΜΑΙΟΣ ΑΡΙΣΤΩΝΟΣ
ΑΜΜΩΝΙΟΣ ΑΜΜΩΝΙΟΥ
ΚΛΕΩΝ ΣΑΚΟΛΑΟΥ
ΔΙΟΝΥΣΙΟΣ ΔΙΟΝΥΣΙΟΥ
ΕΡΜΟΓΕΝΗΣ ΑΓΑΘΙΝΟΥ
ΓΕΝΤΗΚΟΝΤΑΡΧΟΣ
ΑΧΙΛΛΕΥΣ ΔΙΟΝΥΣΙΟΥ
ΑΦΡΟΔΙΣΙΟΣ ΤΙΜΑΓΕΝΟΥ
ΑΣΚΛΗΓΙΑΔΗΣ ΛΙΒΥΣΤΡΑΤΟΥ
ΓΤΟΛΕΜΑΙΟΣ ΘΕΩΝΟΣ
[illegible]ΗΝΑΙΟΣ ΑΘΗ[illegible]Υ
(vacat)

Figure 2. Lefebvre's facsimile copy of the Hermopolis stele (after *Bulletin de la Société royale d'Archéologie d'Alexandrie* 10 (1908) between pp 187-9).

COLUMN I.

<BREAK>
[lines 1-3]
(three names of uncertain interpretation)

[lines 4-18]
With Komanos and the others,
kêrux, Menodoros son of Menodorus,
sêmeiophoros, Theodoros son of Herakleitos,
hekatontarchos, Demetrios son of Apollonios,
(there follows a list of 8 names).

[lines 19-21]
With Atthaniaphas and Sthenelaos,
(2 names)

[lines 22-23]
With Komanos and the others,
(1 name)

[line 24-34]
With Atthaniaphas and Sthenelaos,
sêmeiophoros, Andremon son of Balakros,
pentekontarchos, Dionysios son of Ptolemaios,
(6 names)

[lines 35-49]
And of those formerly with Dryton,
hegemon ep'andron and phrourarch, Timokles son of Timokles,
Officers *exô taxeôn*, (2 names),
Of the *exô taxeôn*, (1 name)
Sons of officers, Dionysios son of Dionysios,
Panteles son of Dionysios,
Diogenes son of Dionysios,
grammateus of the *syntaxis*, Eudoxos son of Timokles.

[lines 50-51]
With Aglaomachos,
Artemidoros son of Antipatros.

[lines 52-55]
From the Thebaïd,
Demetrios son of Antipatros,
Menophilos,
Herakleides son of Leonidas.

[lines 56-57]
Of the Cretans with Aristokartes of Gortyn,
Balakros son of Dionysios.

Figure 3. Simplified listing of the Lefebvre Stele.

[lines 58-59]
Of the Cyrenaicans with Andronikos,
Apollophanes son of Jason.

[lines 60-68]
Of those who have received the amnesty,
archypêretês of the mercenary contingent, Sopatros son of Kassiodoros,
(six other names),

[lines 69-72]
Of the *politikoi* (= citizens/civilians)
(three names).
<BREAK>

COLUMN II.

<BREAK>
[lines 1-65]
(13 names)
pentekontarchos, Hermippos son of Leon,
(4 names)
pentekontarchos, Gytheas son of Neoptolemos,
(31 names)
pentekontarchos, Leonides son of Apollonios,
(6 names)
ouragos, Perigenes son of Alexitimos,

[lines 63-66]
With Diades and his sons,
kêrux, Diophantos son of Jason.
<BREAK>

COLUMN III.

<BREAK>
[lines 1-18]
(9 names)
pentekontarchos, Neon son of Metrodoros,
(5 names)
ouragos, Apollodoros son of Aristomachos.

[lines 19-57]
With Pasines and Dryton,
kêrux, Drakon son of Hegesias,
sêmeiophoros, Mikkos son of Ptolemaios,
hekatontarchos, Ptolemaios son of Tryphon,
pentekontarchos, Zoilos son of Zoilos,
(24 names)
pentekontarchos, Achilleus son of Dionysios,
(4 names)
<BREAK>

and so the inscription most probably dates to shortly after the issue of this amnesty.[19] Numerous other grants of amnesty were made later on during the reign of Euergetes, for example the amnesties granted during the years 121-118 BC to those who had supported Queen Cleopatra and had sided against him in the civil war,[20] and the inscription could possibly be subsequent to any of these in date. The earliest date is the most probable as being closest to the reference to Pasines in the papyrus. If the date of the stele is, in fact *circa* 144 we must perhaps have to assume that the regiment of Cyrenaicans under Andronikos mentioned in col. I, 58 had been sent to Hermopolis by Euergetes when he took the throne late in 145, for it is difficult to see how Philometor could have recruited in Cyrene, given the hostility of the two brothers.

Little chronological evidence can be derived from the names of the other eponymous officers (unit commanders) named in the inscription. The appearance of the personal name Dryton twice in the inscription causes considerable problems of identification. It is quite uncertain whether the Dryton mentioned in col. III, 19 as an eponymous officer along with Pasines is to be identified with the Dryton mentioned in the inscription in col. I, 36 as a former eponymous officer. It is equally uncertain whether either of these individuals can be identified with the well known individual Dryton son of Pamphilos. This third individual called Dryton was born around 195 BC in the city of Ptolemaïs in Upper Egypt and died between 113 to 111 BC.[21] Lesquier thought that the Drytons of the inscription were to be identified with Dryton son of Pamphilos.[22] This is unlikely as Dryton son of Pamphilos was a cavalry officer while the two units in the inscription commanded currently or formerly by a Dryton seem to be infantry units. Furthermore in 153 BC Dryton son of Pamphilos had been transferred to the city of Pathyris over a hundred miles south of Hermopolis. He subsequently remained domiciled in that city apart from postings to the nearby garrison at Krokodilopolis.[23] Any connection seems most unlikely. There seem to have been more than one military officer called Dryton serving in the Ptolemaic army in the middle of the second century.[24] Given the rarity of the personal name Dryton, it is quite possible that these individuals are all related to one another in some way.

Lesquier also thought that the Komanos mentioned as an eponymous unit commander in the inscription could be identified with Komanos of Alabanda, an

---

[19] Winnicki, *op. cit.* p. 13 gives the Lefebvre Stele the same date of 143/2 as the papyrus.

[20] *Ibid.* p. 53.

[21] Jan Krzysztof Winnicki, 'Ein Ptolemäischer Offizer in Thebais' *Eos* 60 (1972) pp. 343-53; Naphtali Lewis, *Greeks in Ptolemaic Egypt* (Oxford 1986) pp. 88-103.

[22] *Rev. Phil* 32 (1908) p. 215.

[23] Naphtali Lewis, *op. cit.* p. 92.

[24] Compare *LGPN* I *s.v.* on the interrelationship of these individuals. Winnicki (*Eos* 60 (1972) p. 347 n. 28) separates Dryton son of Pamphilos from either of the two eponymous commanders called Dryton mentioned in the Lefebvre Stele.

individual who is otherwise attested (*PP* 8559) in a papyrus of around 148 BC as a cleruch farming more than a hundred *aurourai*, and so presumably an officer.[25] This would fit the date suggested above for the Stele very well indeed. Komanos is a quite common name in Ptolemaic Egypt however, and there seems to be no particular reason to associate this regimental commander in the regular army with a farmer of the same name in the army reserve. A simplified version of the list is given as Figure 3.

The most cursory examination of the inscription demonstrates that the old system of infantry organization in *chiliarchiai* and *pentakosiarchiai* and in files of 16 has gone and has been replaced by a completely new system. Logic would suggest that the new sub-unit in the Ptolemaic army, which is clearly based on the Roman maniple, would normally have been divided into two centuries or *hekatontarchiai* each commanded by a *hekatontarchos*, and that each *hekatontarchia* would be divided into two *pentekontarchiai* each being commanded by a *pentekontarchos*.

We do not encounter this logical organization applied in any consistent way. In col. I lines 24 sq. we find a sub-unit from the regiment of Atthaniaphas and Sthenelaos where the standard-bearer is mentioned but not the herald, and he is followed by a *pentekontarchos* and not a *hekatontarchos*. In column II lines 1-65 we find three *pentekontarchiai* commanded by *pentekontarchoi* listed in succession with no mention of a second *hekatontarchos*.

In the case of these irregularities, where no *hekatontarchos* is listed for the sub-unit or only one, or when three *pentakosiarchoi* are listed in succession without mention of a second *hekatontarchos*, it is preferable to assume that these phenomena are to be explained by losses, and by failure to promote successors to these ranks immediately. The general order in which the members of each sub-unit are listed is reasonably standard and can be reconstructed on the basis of Column III lines 19-57 in the following manner:

(Eponymous officer(s) of the unit)
herald,
standard-bearer,

First century.

hekatontarchos,
first *pentekontarchos*,

[25]*Loc. cit.*, cf. Jan Krzysztof Winnicki, *Ptolemäerarmee in Thebais* (Wroclaw 1978) p. 13 n. 13 for other references to a Komanos, perhaps the same individual, ranging in date from 187 to 117.

soldiers of the first *pentekontarchia* of the first century

second *pentekontarchos*,

soldiers of the second *pentekontarchia* of the first century.

Second century.

hekatontarchos,

first *pentekontarchos*,

soldiers of the first *pentekontarchia* of the second century

second *pentekontarchos*,

soldiers of the second *pentekontarchia* of the second century.

ouragos.

We are clearly dealing with a sub-unit organization quite different from that of the earlier Ptolemaic infantry, based on a grouping of two centuries and mentioning titles of rank not previously occurring in the papyri. It seems reasonable to assume that an organizational reform has taken place in the Ptolemaic infantry prior to the erection of the Lefebvre Stele around 144 BC. Furthermore it seems clear, as has already been stated, that the two century organization listed in the inscription is the Ptolemaic equivalent of the Roman maniple.

Section 1 c

The Ptolemaic Maniple and its Officers

In this section we shall first examine the evidence for the Ptolemaic maniple and for the structure of the unit it made up. Then we shall look at the evidence for the different officers of the maniple named in the Lefebvre Stele. The new infantry organizational sub-unit structure outlined in the stele, though not named in that inscription, is evidently to be identified with the *sêmeia*, or 'maniple' which also appears in Ptolemaic papyri for the first time during the second century. The Greek word used for the maniple is 'standard', presumably on account of there being one standard-bearer to each maniple. The same word, spelt in the variant form *sêmaia*, is one of the terms Polybius regularly uses to describe the Roman maniple.[26] It is

[26] See eg. Polyb. 6. 24. 8 where the term *speira* and *sêmaia* are both used of the maniple. Cf. Lesquier p. 93 on Polybian terminology. In the article 'Σημεῖα' in *PW* ii A, 2 (1923) col. 1331-1339

fundamentally important to understand that Greek armies did not use standards at all during the Classical and Hellenistic periods.[27] The fact that the term 'standard' is selected to describe the new type of sub-unit in Greek implies both the introduction of the use of standards, and consequently of Roman tactics, into the Ptolemaic army.

Van't Dack has suggested that the demotic term *stn* is the Egyptian term for the Greek *sêmeia*.[28] The term *stn* is certainly used as an equivalent to *sêmeia* in those examples given by Van't Dack which have a date after the introduction of the *sêmeia* in the 160's, but a number of other examples use the term *stn* before this date (sometimes of cavalry units). It should therefore be assumed that the term is used with the meaning 'military company' and is used of *sêmeia* after the 160s, but is earlier used of *taxis* or of some similar term(s) for an infantry sub-unit. It is possible that *stn* may mean 'standard' in demotic Egyptian of this date, like *sêmeia* in Greek. In such case the word may have come into use in the Achaemenid period, when the Aramaic word for standard, *degel*, was used as the term for a sub-unit in the Achaemenid army in the *lingua franca* of the Empire.

Possibly the first reference to a maniple in the Ptolemaic army comes, not in a papyrus, but in a metrical inscription dedicated to King Ptolemy and King Ptolemy the Brother and Queen Cleopatra by the *prosgraphoi* of the third *sêmeia* of those serving under Hieronymous (οἱ δι' Ἱερωνύμου τοῦ ἀρχ[ισ]ωματοφύλακος καὶ στρατηγοῦ πρόσγραφο[ι] τρίτης σημέας).[29] Hieronymos the *archisômatophylax* and *stratêgos* mentioned in the inscription is the commanding officer of the unit to which the third maniple belonged. The term *prosgraphoi* is not satisfactorily understood, though it perhaps means something like 'added on to the rolls of a unit'.[30] The triple reign of these monarchs began in September 170 and continued, with breaks, down to the second half of 164.[31] The triple reign may have continued to have been cited in some documents for some further time down to the final settlement between the two Ptolemies however. The maniple must therefore have been introduced into the Ptolemaic army before the end of the triple reign.

F. Lammert discusses the use of this word to mean a military signal or naval flag. The discussion is escpecially concerned with its use to denote fire-signals and does not touch on its use to mean maniple.

[27]Johannes Kromayer, Georg Veith, *Heerwesen und Kriegführung der Griechen und Römer* (= edd. Iwan von Müller, Walter Otto, *Handbuch der Altertumswissenschaft IV, 3, 2*, Munich 1928) p. 132.

[28]Van't Dack, *Ptolemaica Selecta* pp. 65-84.

[29] *SB* I no. 1436.

[30] See Lesquier pp. 62-3 n. 3; Launey p. 576 n. 7.

[31] Ptolemy VI Philometor fled Egypt in October 164 effectively ending the triple reign (Edouard Will, *Histoire politique du Monde hellénistique (323 - 30 av, JC)* 2 (Nancy 1982) p. 360; for the chronology cf. Walbank, *Comm.* III p. 322.

As far as the papyrological evidence is concerned the maniple first appears in a document from Memphis dated to 163/2 BC which mentions one Philippos son of Sogenes, a soldier (*stratiôtês*) from the *sêmeia* of Pu[..]rôs (ὁ δὲ Φίλιππος Σωγ-ένου στρατι[ώτ]ης ἐκ τῆς | σημείας τοῦ Πυ..ρῶτος).[32] The damaged name of the eponymous commanding officer is probably best restored as Πυθέρως.[33] This document gives a firm *terminus ante quem* of 163/2 for the introduction of the *sêmeia* into the Ptolemaic infantry. In the Roman army a 'ranker' or private soldier seems to have been termed either *gregarius* or *gregalis*.[34] The fact that Philippos is called a *stratiôtês* in the Memphis papyrus may reflect a change in the official terminology for a ranker.[35] The term *idiôtês* also occurs in the papyri as a title for a private soldier,[36] and all references to this rank come in documents with dates which are earlier, or possibly earlier, than the reforms of the 160's. The evidence is, however, sparse, and it is perhaps best not to read too much into it. The two terms may have been used interchangeably.

An alternative reading of the same papyrus has been suggested which would make Philippos a private soldier in the sixth maniple of Pu[the]rôs (ὁ δὲ Φίλιππος Σωγένου στρατι[ώτ]ης ἑκτῆς | σημείας τοῦ Πυ[θε]ρῶτος). This seems preferable, as the maniples were normally referred to by their number, and the personal name given is that of the regimental officer in charge of a unit of six maniples. For example we find one 'Ptollis standard-bearer of the second *sêmeia*' in a document dated 250-100, which we can now narrow down to a date in the second century BC (*PP* 2388). Apart from dubious readings or interpretations of an 'eleventh' and a 'twenty-first' maniple in the papyri, the highest number attested for any maniple is six.[37] The number of six *sêmeiai* in a unit is confirmed by the Jouget and Roeder stelai discussed below. This unit would be equivalent to a double cohort in the Roman army, although presumably without attached light infantry. This is perhaps the reason why in many cases the infantry unit in the late Ptolemaic army has two eponymous officers rather than one. One supposes that in battle the unit would form up in a *triplex acies* 'chequer' formation two maniples wide and three deep.

The maniple is normally referred to, not by the name of its own commander, the senior *hekatontarchos*, but rather by the name of the unit commander to which the maniple belonged, the eponymous commander. In many cases there were two

---

[32] *UPZ* Vol. I p. 189-192 no. 18 line 5. In line 24 the document is dated to the tenth year of Philometor, see p. 188.

[33] The name occurs in an inscription recording victors in the Erotideia at Thespiai (see Louis Robert, *Hellenica II* (Paris 1946) p. 6 line 22). I owe this information to Richard Catling.

[34] Yann Le Bohec, *The Imperial Roman Army* (Oxford 1994) p. 47.

[35] Thus the Greek word *stratiôtês* is used of a soldier of the XII legion in an Imperial period incription from Hadrianutherai (*IGR* iv 233).

[36] *PP* 3813 (second century), 3906 (257/6), 3904a (c. 270-50), 3920 (before 210, though the reading of *idiôtês* in this text is uncertain).

[37] R.W. Daniel, 'The Military σημεῖον in P.Amh. II 39' *ZPE* 52 (1983) pp. 269-71.

eponymous commanders of the unit. The maniple can be described in two ways. Either the maniples are cited as 'the first to sixth maniple of (the unit commanded by) one or two eponymous officers', or more simply as 'a maniple of (the unit commanded by) one or two eponymous officers' but without any number. For example columns II and III of the Lefebvre Stele apparently list maniples in series, one after another. The eponymous officers named there, frequently in pairs, are not maniple commanders but rather the eponymous commanders of the six-maniple unit as a whole. For this reason in column III, 19 sq. after the title 'Under Pasines and Dryton', the first name to be listed, that of the principal hekatontarch of the maniple, is Ptolemaios son of Tryphon, not Pasines or Dryton which we would have expected if the eponymous officers commanded the maniple rather than the unit.

It seems that under certain circumstances, perhaps only in certain units, it was the centuries which were numbered instead of the maniples. A unique document, an undated papyrus from Tebtunis, mentions a soldier 'from the troops commanded by Polykrates, of the 8th century of the Macedonian *agêma*'.[38] The interpretation of the numeric symbols in this text is perhaps open to dispute, but it would be difficult to postulate a more likely one.[39] If they are correctly interpreted, we have a unique example of numeration by centuries rather than maniples. Perhaps the phenomenon of numbering by centuries was confined to guard units, which this unit presumably is, given the fact that it is referred to as both the *agêma*, a term meaning 'vanguard' which was only given to élite units in the Hellenistic armies, and as Macedonian. This same unit, the *agêma*, is mentioned in Polybius' account of the battle of Raphia in 217 BC, where it numbered 3,000 men.[40]

Having examined those documents mentioning the *sêmeia*, we have been able to demonstrate that the *sêmeia* was introduced into the Ptolemaic infantry before 163/2, and that six *sêmeiai* made up an infantry regiment. We shall now turn to examine the references to the officers of the *sêmeia* listed in the Lefebvre Stele in other documentary sources, firstly to see whether all dated documents referring to these ranks confirm our *terminus ante quem*, and secondly to see if they can throw any further light on the internal organization of the *sêmeia*.

The Lefebvre Stele demonstrates that the *sêmeia* was divided into two centuries, presumably called *hekatontarchiai*, commanded by *hekatontarchoi* 'commanders of a hundred'. No *hekatontarchos* attested in the papyri (*PP* 2321-2232) has a date earlier than the second half of the second century BC.[41] Normally the rank of hekatontarch was denoted by a palaeographic symbol. It seems that Bouché-Leclercq was the first to note that in papyri of the Imperial period the rank of centurion in the Roman army was denoted by the palaeographic sign *rho* (ρ́), which

---

[38] *SB* I 4318, 2 τῶν Πολυκράτου τῆς η ρ ἀγή(ματος) Μακεβονικῶν.

[39] Lesquier p. 80.

[40] Polyb. 5. 65. 2; G. T. Griffith, *The Mercenaries of the Hellenistic World* (Cambridge 1935) p. 119; Cf. Lesquier pp. 11-12, 21-24.

stands for the number 100 in the Greek alphabetic system of numeration, usually with the addition of a cross above the letter.[42]

Some time after Bouché-Leclercq had made this observation a papyrus from Diospolis Magna was found recording an agreement made on 26 June 119 BC which mentions one Sesoôsis. The name is a variant version of the personal name Sesostris. When Sesoôsis is mentioned his name is followed by the palaeographic symbol ρ́. In interpreting this Ptolemaic document both Lesquier and Wilcken suggested that the same palaeographic symbol must stand for centurion, or *hekatontarchos*, in documents of the Hellenistic period, despite their early date.[43] The earliest holder of this rank seems to be one Noumenios, attested as Νουμήν[ι]ος ρ́ in a papyrus from Tebtunis in the Fayoum, a fragmentary list of Greek and Jewish settlers and cultivators giving acreages and rentals, dated by its editors to the mid-second century BC.[44] The editors tentatively accepted that an interpretation of the palaeographic symbol following the name of Noumenios as signifying hekatontarch is 'possible'. The current state of our knowledge perhaps transfers this possibility into a certainty.

The Lefebvre Stele further demonstrates that in the Ptolemaic maniple there was at least one and more probably two *pentekontarchoi* or 'commanders of fifty' serving under each *hekatontarchos*. They presumably commanded a sub-unit termed a *pentekontarchia*. The earliest reference to a Ptolemaic pentekontarch comes in a letter of 162 BC from Tebtunis in the Fayoum, containing a reference to the price of wheat, which is valued at 500 copper drachmae per artaba, and mentioning [---]ν τῶν Πολυκράτου πεντηκόνταρχος.[45] All other references to *pentekontarchoi* gathered in *Prosopographia Ptolemaica* (2333-2366) are later except for *PP* 2354, an unedited papyrus of 243/2 BC from Herakleopolis mentioning an individual called Neo[---]who holds the abbreviated rank of *[p]en(tekontarchos)* in the regiment of Nautas (Νέο[---] [π]εν(τηκόνταρχος) τῶν Ναυτᾶ). In interpreting this document the abbreviation *pen-* has been expanded into *pen(tekontarchos)* rather than *pen(takosiarchos)* quite arbitrarily, even though both interpretations are equally possible. It is but one example of a number of cases in the papyri where a personal name is followed by this abbreviation.[46] Clearly the

---

[41] *PP* 2324 is given a date of ±250/100. The hekatontarch has the Egyptian name Thusutmês son of Thusutmês and so is likely to be later rather than earlier in date. Similarly *PP* 2331, mentioned in the papyrus *PBad.* IV 84, 47, which was dated to the second century by the editors, likewise has the Egyptian name Horos, and has the general appearance of being later rather than earlier in date.

[42] A. Bouché-Leclercq, *Histoire des Lagides IV* (Paris 1907) p. 225.

[43] Lesquier p. 160 n. 3; Wilcken, *UPZ* Vol. II no. 161 (Πολυπέρχοντους | τοῦ 'Αμμωνίου τῶν κατοίκων ἱππέων, Σεσοώσιος ρ́ φιλοβασιλιστῶν προθύμων καὶ ἄλλων), commentary on p. 56 = *PP* 2326.

[44] *PTebt.* III, ii p. 218 no. 1019 = *PP* 2326.

[45] *PTebt.* III, ii p. 191 no. 910 = *PP* 2362.

[46] *PP* 2342, 2343, 2354, 2357, 2358, 2360, 2361, 2363.

abbreviation should be restored as *pen(takosiarchos)* where it occurs in documents before the 160's. The *pentakosiarchos* was the commanding officer of an infantry unit of 500 men, derived from the Alexandrine *lochos*, into which the Ptolemaic infantry was organized throughout the third century and down to the infantry reform of the 160's. When the rank of the *pentekontarchos* is given by a palaeographic symbol, as with the *hekatontarchos*, the alphabetic Greek number for 50 is given, in this case the letter *nu* (ν́).

The rank of a commander of fifty is not attested for the Roman army in any documents, literary or epigraphic, from the Republican period. Isidore of Seville, writing in the seventh century AD mentions *quinquagenarii, quia in capite sunt quinquaginta militum*.[47] It is difficult to know whether this may be genuine material relating to the early organization of the Roman army or not. Livy (8. 8. 18) mentions that during the Latin War (340-338 BC) the chief centurion (*primus pilus*) of the Roman army obtained permission from the consuls to chose whom he liked for his *subcenturio*. Livy may be alluding to the *quinquagenarii* in the passage, although there are other possibilities. Normally the senior of the two centurions in each maniple (*centuriones priores*) were selected first, and then the junior centurions (*centuriones posteriores*) for each maniple.[48] It is perhaps preferable to suppose that Livy could be telling us that the *primus pilus* received permission to chose his own junior centurion for the first maniple of the *triarii*.[49] The existence of a commander of fifty in the late Ptolemaic infantry is certain, but it seems unlikely that this represents contemporary Roman practice. Rather it is a refinement of Roman practice.

It is perhaps worth noting that evidence also exists for the rank of commander of fifty in the Jewish army which was formed by Judas Maccabaeus. This army was formed in 165 BC, at exactly the time when the Seleucid and Ptolemaic armies were undergoing reform along Roman lines. *1 Maccabees* 3. 55 tells us that Judas appointed commanders of thousands, hundreds, fifties and tens. These ranks, including that of commander of fifty, may have been introduced under the influence of existing practice in the Seleucid or Ptolemaic army. It is equally possible, however, that they were introduced under the influence of Jewish religious writing: *Deut.* 1. 15 mentions Moses as appointing the same officers.[50] The Dead Sea Scroll known as 'The War of the Sons of Light Against the Sons of Darkness' likewise describes the organization of an imaginary army of the Sons of Light organized into thousands, hundreds, fifties and tens. The date and authorship of this document are disputed, but it is generally thought to have been written in the second half of the first century BC by a member of the Essene sect. The organization of this imagi-

[47] Isidore of Seville, *Etymologiarum* 9. 3. 31.

[48] The process is described by Polybius (6. 24. 1)

[49] Cf. Nick Sekunda, Richard Hook, *Republican Roman Army 200-104 BC* (London 1996) p. 18.

[50] Cf. Josephus, *AJ* 12. 7. 3, who does not mention the last two ranks and who states that the organization was carried out according to ancestral custom.

nary army may well be based on practice in late Hellenistic Jewish armies.[51]

Other than this the *pentekontarchia* is also referred to as a sub-unit of 64 light infantry in the so-called Greek *Taktika* of Asclepiodotus (6. 3), Aelian (*Tact.* 16) and Arrian (*Tact.* 14), which all seem to be ultimately dependent on a work of Poseidonius and therefore describe an 'ideal' army in the early first century BC. The argument for this view is presented in Appendix B. Each *pentekontarchia* would presumably have been commanded by a *pentekontarchos*, and two of these *pentekontarchiai* made up a *hekatontarchia*. It seems to be a reasonable assumption that the lost work of Poseidonius upon which all of these works were ultimately based, drew upon contemporary late Hellenistic Ptolemaic or Seleucid practice for the title of *pentekontarchia*.

No subordinate officers are attested below the rank of pentekontarch, but it is possible that the *penkekontarchia* was divided into a number of tent-parties.

Having established that the Ptolemaic maniple consisted of two centuries, each broken down into two *pentekontarchiai*, we now need to make an attempt to estimate the strength of the Ptolemaic maniple, using the Lefebvre Stele as our guide. As has already been mentioned, we do not understand the precise nature of the stele. Again, one would be justified in taking the Stele as recording the parade state of some of these *sêmeiai* whilst performing garrison duty in a relatively remote posting, as opposed to their establishment strength. By the latter term we mean the theoretical size of a unit or sub-unit which the military authorities would wish to maintain under ideal circumstances. Normally the parade state of the unit will fall considerably below the establishment strength. By this second term we mean the number of men who are actually present when the unit carries out its daily muster parade. This figure excludes dead, wounded, ill, persons detached from the unit and initial failures to recruit up to the establishment strength. In order to make a calculation as to the strength of the Ptolemaic maniple we shall assume that the Stele lists all the members of the sub-units present, that is their parade state.

In column II we can see *pentekontarchiai* with actual 'parade states' of 13+ for the first *pentekontarchia* of the first *hekatontarchia* and 5 for the second. In the second *hekatontarchia* of the *sêmeia* the two *pentekontarchiai* have strengths of 32 and 7. In the *sêmeiai* listed in column III the one at the top of the column has a second *hekatontarchia* with *pentekontarchiai* numbering 10+ and 6, while the second *sêmeia* has *pentekontarchiai* of 25 and 5+ in its first *hekatontarchia*. These *sêmeiai* had, therefore, fallen well below their establishment strengths due to transfers, death and disease, and discharge upon termination of contract. Under these circumstances the practice seems to have been to maintain the first *pentekontarchia* of each *hekatontarchia* at a strength sufficient to preserve its operational

---

[51] Yigael Yadin, *The Scroll of the War of the Sons of Light against the Sons of Darkness* (Oxford 1962) p. 59 (organization), 243-6 (date), 246 (authorship).

capacity, but to allow the second to fall to a cadre strength. The maximum strength given to any *pentekontarchia* in the document is 32 (i.e. 31 plus the pentekontarch).[52] This figure, then, must constitute our starting point in any calculation of the strength of the Ptolemaic maniple, which would therefore have had a minimum establishment state of 128 officers and men.

We can compare this minimum figure with contemporary evidence for the Roman army. Polybius (6. 21. 9) states that there were normally 60 men in a maniple of *triarii* and 120 men in a maniple of *principes* or *hastati*. The size of the Roman maniple, and therefore of the Roman legion as a whole, was increasing rapidly during the first decades of the second century however. By the 160's the maniple may have been approaching a strength of two hundred men. For example Livy (33. 1. 2) notes that in 197 Flamininus had 2,000 *hastati* in his legions, in other words 200 per maniple.

The only other comparison available is the maniple as described in the 'ideal' army of the Greek *Taktika*. We have already mentioned that the *Taktika* mention a *pentekontarchia* of 64 men in the light infantry, but in the heavy infantry this sub-unit is called the *tetrarchia*. Thus Asclepiodotus (2. 8) writes that in the heavy infantry two *tetrarchiai* of 64 men each commanded by a *tetrarchos* made up a *taxis* of 128 men commanded by an officer who used to be called a *taxiarchos*, but who is 'nowadays' known as a *hekatontarchês*, or centurion. It is interesting to compare this usage with that of Polybius (6. 24) who uses the terms *taxis* and *taxiarchos* for century and centurion in the Roman Army. The Asclepiodotan rank of *tetrarchos* was probably based on the Antigonid army, for which the rank of *tetrarchos* as a commander of a unit of 64 men is attested in an inscription dating to the reign of Philip V.[53] In the Asclepiodotan system two centuries made up a *syntagma*, a maniple, of 256 men.

In summary then almost any figure between 128 and 256 seems possible for the establishment strength of the Ptolemaic maniple. The best way of proceeding further would be to look at the lowest sub-unit in the maniple. In Asclepiodotus (2. 2) and the other *Taktika* this is the file of 16 men, which he calls a *lochos*. The choice of 16 men need not necessarily follow contemporary late Hellenistic practice. It may rather have appealed to the original compiler, perhaps Poseidonius, because of the fact that the figure can divide down to unity, that is divide by two down to one, which other figures cannot, and because it was the strength of the *dekas* in Alexander's army and of the files in most Hellenistic armies which followed it. A figure of 256 for the establishment strength of the Ptolemaic *sêmaia*

---

[52] Noted by Jan Krzysztof Winnicki, *Ptolemäerarmee in Thebais* (Wroclaw 1978) p. 13.

[53] P. Roussel, 'Un Règlement militaire de l'Époque macédonienne' *RA* 1934, 1 pp. 39-47; Gaetano De Sanctis, 'Epigraphica XII. Il regolamento militare dei Macedoni' *Riv. Fil.* 12 (1934) pp. 515-21; Michel Feyel, 'Le Règlement militaire trouvé a Amphipolis' *RA* 1935, 2 pp. 29-68, who merely refers to previous work on p. 45; F. W. Walbank, *Philip V of Macedon* (Cambridge 1940) p. 293.

therefore seems unlikely.[54]

There is some evidence that the basic building-block of the Roman maniple and century was the tent-party (*contuburnium*) of 6 men,[55] so the century might be expected to rise in multiples of 6. If we take the figure of 32 as a minimum for the Ptolemaic *pentekontarchia*, it is plausible that its establishment strength may have been 36 (six tent-parties), giving a total strength of 144 men (perhaps plus 4 officers) to the maniple, or 42 (seven tent-parties), giving a total strength of 168 men (perhaps plus 4 officers), or 48 (eight tent-parties), giving a total strength of 192 and so on. Of all these configurations the most preferable would be the lowest alternative: six tent-parties of 36 men. As mentioned earlier, the *pentekontarchiai* listed in the Lefebvre Stele are well under-strength, though some attempt seems to have been made to keep the first *pentekontarchia* in the century at near establishment strength. If the establishment strength of the *pentekontarchia* was any higher than 32 given a half-file of 8, or 36 given a tent-party of 6, the *pentekosiarchiai* of the *sêmaiai* stationed at Hermopolis were falling quite severely below strength.

Section 1 d

The Under-Officers of the Ptolemaic Maniple

Having concluded our examination of the *sêmaia*, of its officers and sub-units, we shall now turn to examine its 'under-officers'. First of all we shall establish how many under-officers were found in the Roman Republican maniple. Then we shall examine the evidence for what under-officers were found in the Ptolemaic *sêmaia*, using both the evidence of the Lefebvre Stele and other documentary sources. Finally we shall attempt to establish what this corps of under-officers was called in official Ptolemaic military terminology.

Aside from the officers who commanded it, the Roman maniple also included a number of individuals who were not included in the ranks. In later times the Latin

---

[54] Cf. Winnicki, *op. cit.* p. 13 'wahrscheinlich je 250 Personen zählenden Hegemonie'.

[55] For example Josephus *BJ* 3. 124, 5. 48 tells us that in the first century AD the Roman army marched six abreast and Schulten's excavations of the Roman camp at Nobilior in Spain where the century, numbering 60 men according to Polybius, camped in ten *contubernia*. The evidence is gathered and discussed by Everett L. Wheeler, *Chiron* 9 (1979) p. 312. In Appendix B I argue that the Sidon tombstones belong to soldiers of the Ptolemaic garrison stationed there in the years 147-5. One of these (*Rev. arch.* 1904, 2 pp. 8-9 no. 5) was erected in memory of Aristeidas son of Aristeidas a Lakedaimonian from Gytheion by Alexon and Tetartidas his φίλ[οι καὶ σύ]σκηνοι. The second word may perhaps mean that they were members of the same *contuburnium*. It should be noted though, that the word *syskênoi* has a long history in the Greek language.

term for this 'staff' of the maniple was *principales*, a term distinguishing them from the private soldiers in the ranks.[56] For the sake of convenience we shall call this 'staff' the under-officers of the maniple. We run into considerable difficulties when trying to establish the composition of the *principales* of a maniple in the Roman army during the early second century BC. Polybius deals with them in passing. He uses Greek terms for them, and it is sometimes unclear how these correspond to their Latin counterparts. The composition of the *principales* in the Imperial army is known in greater detail, but it is uncertain how much of this information can be transposed back to the Republican period. The number of *principales* grew during the Imperial period, for example Vegetius (1. 7) mentions a long list of *principales*, but this list refers to fourth-century AD conditions.[57] Furthermore many of the posts mentioned by Vegetius belong to levels of organization higher than the manriple.

Polybius (6. 24. 2) notes that when a legion was formed 60 centurions (*taxiarchoi*) were elected first of all and then each centurion appointed an *ouragos* (a 'file-closer').[58] It is quite certain that Polybius uses this term for the Latin *optio*.[59] In the Imperial Roman army the *optio* carried a large staff with which to beat men back into the ranks. He probably stood behind the century on the left-hand side.[60] It is certain, therefore, that the rank of *optio* already existed in the early second century BC.

After completing his description of the manner in which the officers of the legion were selected, Polybius (6. 24. 6) goes on to tell us that two standard-bearers, *sêmaiophoroi* in Polybius' vocabulary, were selected from each of the maniples. It is difficult to accept his statement, for although each century had its own standard-bearer in the Imperial period, it seems that in the Republican period it was each maniple and not each century that had its own standard-bearer. Varro (*Ling.* 5. 88) writes that each maniple had one standard (*manipulos exercitus minimas manus quae unum secuntur signum*). Varro published his *De lingua latina* before the death of Cicero, and so was writing in the 40s of the first century BC.[61] Similarly Lucan (1. 296), writing in the first century AD, tells us that during the Civil Wars Caesar summoned his armed maniples to the standards (*convocat armatos*

[56] David J. Breeze, 'The Career Structure below the Centurionate during the Principate' in *Aufstieg und Niedergang der Römischen Welt II. 1* (Berlin 1974) pp. 435-51 notes (p. 449) that the term *principalis* is first attested in AD 107, however the term may well have been in existence in Republican times too.

[57] G. R. Watson, *The Roman Soldier* (London 1969) p. 78.

[58] Cf. Festus s.v. *Optio.*

[59] Walbank, *Commentary* I p. 707. Varro (*Ling.* 5. 91) tells us that in his day the tribunes appointed the *optiones*. Cf. Festus (*s.v.*) 'Optio est optatio, sed in re militari optio appellatur is, quem decurio aut centurio optat' etc.

[60] Michael Speidel, *The Framework of an Imperial Legion* (The Fifth Annual Caerleon Lecture, Cardiff 1992) 24-5.

[61] *OCD*^{2} p. 1108.

extemplo ad signa maniplos), implying that the standards were assigned to the maniples rather than to the centuries. These sources are, however, considerably later than Polybius, who could be correct if there was a change in practice in the later second century. Walbank gets round the problem by suggesting that there was one standard but two standard-bearers per maniple.[62]

Polybius (6. 34. 8) finally mentions that one man in each maniple was made responsible for passing on the password and that he was relieved from the chore of performing guard-duty. This soldier must correspond to the *tesserarius* of the Roman Imperial Army, who was responsible for selecting small sentry pickets and fatigue parties as well as for passing on the password.[63] Later on Polybius (6. 36) describes how the guards of each maniple in the legionary camp were checked and their *tesserae* (tokens of the watch) were collected by patrols formed by the cavalry of the legion.

It is generally held that in the Imperial period the *principales* at the level of century were limited to a 'tightly knit' group of three: the *signifer*, the *optio* and the *tesserarius*.[64] It is reasonable to assume that these three offices also existed in the Roman Republican army, though perhaps originally at the level of maniple rather than century. Breeze notes that it was precisely these three *principales* who are mentioned by Polybius.[65] The relative level of importance of these three ranks was, in ascending order, *tesserarius*, *optio* and then *signifer* at the top.[66]

The conclusion of our discussion of the Roman evidence therefore allows us to conclude that it was these three under-officers who were found in the Roman Republican maniple. We shall now turn to examine the Ptolemaic evidence. We would also expect to find the same three under-officers in the Ptolemaic *sêmaia*.

The Lefebvre Stele is of great value in enabling us to decide which of the under-officers mentioned in the Ptolemaic papyri are found at *sêmaia* level and which are found higher. It confirms that exactly the same three under-officers are found at sub-unit level in the Ptolemaic army as in the Roman Republican army as described by Polybius. When the stele lists a *sêmaia*, first, at the top of the listing comes a *kêrux* or herald, equivalent to the *tesserarius*, and then a *sêmeiophoros* or standard-bearer, the equivalent of the *signifer*, then come the officers and men of the *sêmaia*, and then finally, at the very bottom, comes an *ouragos* or 'file-closer', the equivalent of the *optio*. These positions in the list presumably mirror their positions in the line of battle. Like the *optio* the *ouragos* would stand behind the *sêmeia* so as to ensure no-one left the ranks, and for this reason he is listed at the end of the *sêmeia*.

---

[62] Walbank, *Commentary* I p. 707 ad. loc.

[63] On the *tessararius* see Graham Webster, *The Roman Imperial Army*^{2} (1979) p. 120.

[64] G. R. Watson, *The Roman Soldier* (London 1969) p. 79.

[65] Breeze, *op. cit.* esp. p. 448 n. 69 'Polybius records the existence of the posts of *tessararius*, *optio* and *signifer*'.

[66] Breeze, *op. cit.* p. 442.

The earliest attested occurrence of the terms *kêrux* (*PP* 2390-2399) and *stratokêrux* 'army-herald' comes in a papyrus from Diospolis Magna dated 1 January 162 mentioning one Αρχέλαος κῆρυξ τῶ[ν σ]τρατοκηρύκων.[67] It is difficult to believe that the Ptolemaic army existed without heralds before the 160's however, so the absence of them in the records is perhaps a reflection of the sparsity of the evidence rather than of the non-existence of the rank before the 160's.

The earliest of the standard-bearers (*PP* 2379-89) is Korax son of Dionysios (*PP* 2385) standard-bearer of those troops under Pasines,[68] attested in a papyrus probably from Hermoupolis Magna, which has already been mentioned with reference to the date of the Lefebvre Stele, and which dates to 143/2 BC: a date later than the reforms of the 160's. Korax's rank is spelt *sêmeophoros* (σημεοφόρος) in this document, instead of the *sêmeiophoros* we might expect. A further variant spelling of *sêmophoros* (σημοφόρος) is found in a metrical inscription from Koptos, where the word has possibly been abbreviated for metrical reasons.[69]

The situation concerning the rank of the file-closer, or *ouragos*, is not at all clear. All the *ouragoi* listed in *Prosopographica Ptolemaica* date to the second half of the second century or to the first, except for one (*PP* 2369) which dates to 248/7 BC. The identification of this individual as an *ouragos* depends on the interpretation of a palaeographic symbol combining the letters *omicron* and *upsilon* as standing for that rank. The document in question concerns regulations for the maintenance of cavalry horses. It contains a list of personal names, reproduced here as Figure 4, each with a palaeographic symbol on the left side. In five cases the name and symbol are further qualified by an ου symbol still further to the left. In no case does the ου symbol stand by itself opposite a name. In the penultimate incomplete line above the list of names the phrase 'by the horse-inspectors' (ἱπποσκοποῖς) can be read. It is evident, therefore, that the list was compiled during a horse-inspection. The original publishers of this papyrus, Mahaffy and Smyly, suggested that all the palaeographic symbols stood for ranks, and that the ου symbol 'may perhaps represent ουραγιας or ουραγος'.[70] This would entail that five of the individuals in the list held two ranks simultaneously, which is clearly impossible. Smyly subsequently suggested that the ου symbol is more likely to stand for a negative,[71] i.e.. that the individual in the list either had not had his horse inspected, or that the

---

[67] *BGU* III 992, ii line 4 = *PP* 2390.

[68] *BGU* VI 1258a lines 15-16 = *PP* 2385. For the name of the eponymous commander of the unit, see Jan Krzysztof Winnicki, *Ptolemäerarmee in Thebais* (Wroclaw 1978) p. 13.

[69] Listed as *PP* 2386. Original publication by A. Wilhelm, Αἰγυπτιακά (*Sitz. Akad. Wien,* Vienna 1946) pp. 38-40 no. 11 lines 3-4; Werner Peek, *Griechische Vers-Inschriften I* (Berlin 1955) pp. 328-9 no. 1149 dates this inscription to the beginning of the second century, but it certainly post-dates the reform. Also published in É. Bernand, *Inscriptions métriques de l'Égypte gréco-romaine* (Paris 1969) pp. 49-52.

[70] *PPetrie* III p. 158 no. 52 (a) 4 Col II, note to line 4.

[71] J. Gilbart Smyly, *Greek Papyri from Gurob* (1921) p. 32 no. 18.

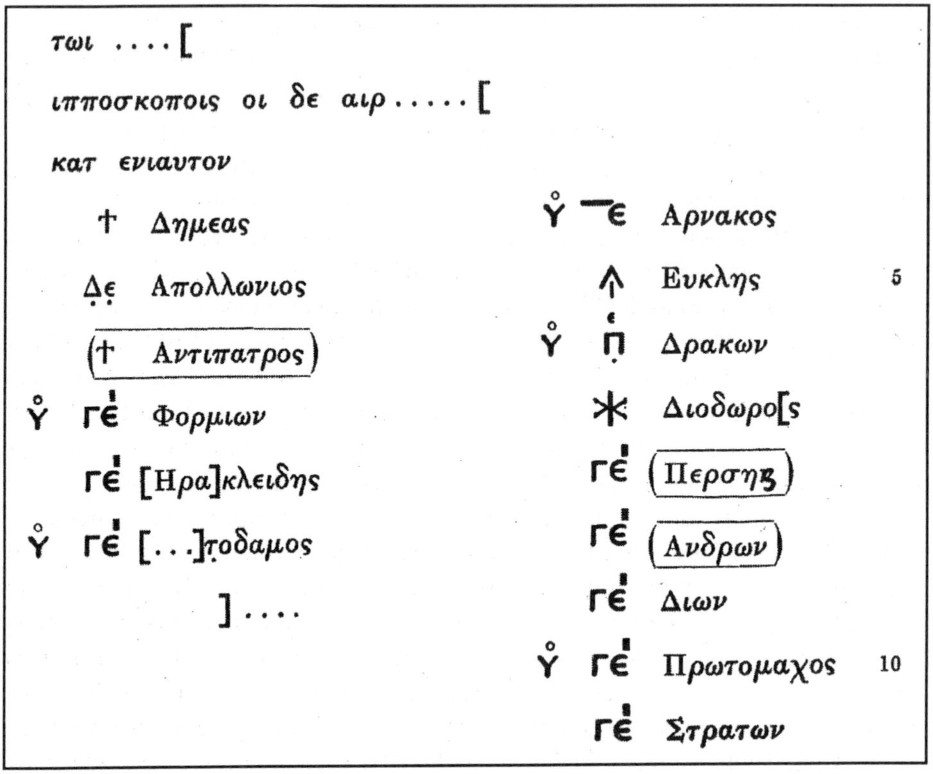
τωι ....[

ιπποσκοποις οι δε αιρ.....[

κατ ενιαυτον

† Δημεας

Δε Απολλωνιος

(† Αντιπατρος)

γε Φορμιων

γε [Ηρα]κλειδης

γε [...]ṭοδαμος

]....

ε Αρνακος

Ευκλης

π Δρακων

Διοδωρο[ς

γε (Περσης)

γε (Ανδρων)

γε Διων

γε Πρωτομαχος

γε Στρατων

Figure 4. Reproduction of majuscule in *PPetrie* III p. 158 no. 52 (a) 4 Col II.

horse had failed the inspection. This suggestion was accepted by Uebel.[72] The individuals from this list are probably therefore not *ouragoi* at all, and should be removed from their listing as such in *Prosopographica Ptolemaica*. It is possible, then, that the rank of *ouragos* too only appeared in the 160s.

In a dossier of correspondence relevant to an Egyptian *sêmaia* commanded by Pates and Pachrates during the Palestinian war of 103-1 BC,[73] we find references to Horos son of Portis the standard-bearer, and to Horos son of Nechoutes, 'the man who has been elected'. This second Horos is known to have been an older man, probably in his 40s. The title he is given looks like a semantic translation into demotic Egyptian of the Latin term *optio*. Clearly Horos son of Nechoutes holds the rank of *ouragos*. Pentekontarchs are nowhere mentioned in these docu-

---

[72] Fritz Uebel, *Die kleruchen Ägyptens unter den ersten sechs Ptolemäern* (Berlin 1968) p. 205 n.4.

[73] E. Van't Dack, W.Clarysse, G.Cohen, J. Quaegebeur & J.K. Winnicki, *The Judean-Syrian-Egyptian Conflict of 103-101 B.C.. A Multilingual Dossier Concerning a "War of Sceptres"* (*Collectanea Hellenistica I,* Louvain 1989) pp. 37-81.

ments, as they are in the Lefebvre Stele.

Having established that the Ptolemaic *sêmaia* had three under-officers, the herald, the standard-bearer and the file-closer, we shall now attempt to establish what the collective title for these various grades of under-officer was in official Ptolemaic military terminology. We shall first of all return to discuss the situation as described by the writers of the Greek *Taktika*. As mentioned earlier these handbooks describe an army organized along Roman lines and divided into maniples, a maniple there being called a *syntagma*, of 256 men. As well as the officers of the maniple Asclepiodotus mentions the military herald (*stratokêrux*), standard-bearer (*sêmeiophoros*), trumpeter, attendant (*hypêretês*), and file-closer (*ouragos*) as the under-officers also belonging to its staff. Asclepiodotus attempts to explain the tasks of these five under-officers most unconvincingly. He tells us that the *stratokêrux* passed on what had been ordered by voice, the *sêmeiophoros* by a signal if the voice could not be heard, then the trumpeter passed on the order by trumpet if the signal could not be seen. Asclepiodotus adds that the *hypêretês* was there to bring along whatever was needed and the *ouragos* was there to bring up anyone who had left the ranks. All this information seems to be a misunderstanding brought into the text by Asclepiodotus. It is not given in the corresponding passages in Aelian (9) or Arrian (10) and it was presumably not found in the original text of Poseidonius. I believe that it was the intention of Poseidonius to mention all under-officers at the level of the maniple even though some, such as the *hypêretês* who will be discussed later, were found at a higher, regimental level. The trumpeter is not mentioned at all in the Ptolemaic documentary evidence.

The term given to the 'under-officers' of the maniple by Asclepiodotus (2. 9) is *ektaktoi* (ἐκτάκτοι). Asclepiodotus explains the etymology of the term in that they were formerly attached to the century (*taxis*) for 'they were not counted as belonging to the *taxis*': now they were attached to the maniple instead. This seems to be an etymological blunder, presumably added to the original text of Poseidonius by Asclepiodotus himself. One notes, once again, that this information is not given in the corresponding passages in Aelian (9) or Arrian (10). In fact the term *ektaktoi* is better understood as meaning 'those drawn up outside', as they were not included in the fighting ranks of the maniple but stood in front or behind them. It is therefore preferable to believe that these *ektaktoi* were attached to the maniple, not the century, right from the beginning. The Greeks did not have a universal term for those holding rank between officers (*hêgemones*) and private soldiers (*stratiôtai/ idiôtai*), but it is presumably Poseidonius who first included the term *ektaktoi* in his original text, a term which is drawn from contemporary late Hellenistic practice.

---

[74] See the detailed treatments in Maurice Holleaux,''Ηγεμὼν τῶν ἔξω τάξεων' in *Études d'Épigraphie et d'Histoire grecques III Lagides et Séleucides* (Paris 1942) pp. 1-14 and in Van't Dack, *Ptolemaica Selecta* pp. 65-84. The traditional opinion used to be that the term was used of soldiers on the reserve rather than the active list (see e.g. T. B. Mitford in *Opus. Athen.* 1 (1953) p. 153).

The phrase *hoi exô taxeôn* (οἱ ἔξω τάξεων) and related terms appear in Ptolemaic documents, largely dating after the 160's. I believe that in Ptolemaic usage after the infantry reform of the 160's this term was used as the equivalent term for the Latin *principales*, meaning 'those outside the ranks', and corresponding to the *ektaktoi* in the *Taktika* of Asclepiodotus and other writers. Current explanations of this and related Ptolemaic terms are not entirely satisfactory.[74]

The term first appears in passages of Diodorus dealing with the reign of Alexander and its immediate aftermath. Diodorus 17. 83. 2 notes that Alexander settled the cities he founded about Alexandria in the Paropomisadai with, among others τῶν ἐκτὸς τάξεων. In the parallel passage in Arrian (4. 22. 5) this group is described as ἀπόμαχοι or 'those unfit for fighting'. Cohen used this pair of passages, supported by the following passages in Diodorus, to suggest that the phrase was used of invalided soldiers who were no longer fit for service in the ranks.[75] The two passages are not precisely parallel however. In Diodorus, Alexander settles a number of cities one day's march from Alexandria, but in Arrian it is the city of Alexandria itself which is being settled. The two texts clearly do not offer themselves to be used as mirror images as Cohen has treated them. In all probability the term *hoi exô/ektos taxeôn* has a more general meaning and is given to those accompanying the army as traders etc. and not enrolled in the ranks. The phrase is repeated by Diodorus (19. 22. 2) during his description of a feast organized by Peucestas in 317 for the army of Eumenes in Persis, and is also used by him (19. 49. 3) of that part of an army which did not receive rations during the siege of Pydna in 316. Finally Diodorus 19. 81. 4 tells us that even those ἐκτὸς τάξεως would flock to hear the young Demetrius Poliorcetes speak.

The term passed into Ptolemaic usage. An inscription has preserved a letter, possibly of Ptolemy IV to a provincial governor, ordering relief from billeting for the city of Soloi. In this letter mention is made of the *exô taxeôn*, and from the context it seems clear that these are non-combatants or camp-followers.[76] Thus down through the third century BC it retains the same meaning in Ptolemaic military terminology as it had in the age of Alexander. After the infantry reform of the 160's, however, the term is probably best understood as meaning the 'under-officers' who stood outside the ranks of the *sêmeia*.

---

[75] D. Cohen, 'Annotationes ad Auctores et Papyros Nonnullas II οἱ ἔξω τάξεων' *Mnemosyne* 54 (1926) pp. 82-3.

[76] C. Bradford Welles, *Royal Correspondence in the Hellenistic Period. A Study in Greek Epigraphy* (London 1934) p. 138.

Section 1 e

The Ptolemaic Infantry Regiment

We have now concluded our discussion of the Ptolemaic *sêmaia*, having established that six of these *sêmaia* made up a unit, or regiment, in the Ptolemaic infantry. We shall now move from the level of the sub-unit, the maniple, to the level of the regiment. In this section we shall first discuss the organization of the regimental headquarters, and in particular which officers and under-officers were found there and what there tasks were, and then we shall attempt to establish what this infantry regiment was called in official Ptolemaic military terminology.

Once again the Lefebvre Stele shall be our guide. The items listed in the first column of the stele are very mixed. Individual citizens not belonging to the military are mentioned,[77] and odd individuals from a variety of military units who happen to be present at the fort at Hermopolis Magna. Nevertheless some interesting conclusions can be drawn from Column I lines 35-49, which would seem to list the headquarters element of the unit called 'Those formerly with Dryton'.

The first individual listed is the current commander of the unit Timokles son of Timokles,[78] who, one presumes, has recently replaced Dryton in command, although the official title of the unit has not yet changed. As we have seen, the Stele was inscribed shortly after the death of Ptolemy IV and the seizure of power by Ptolemy VIII Euergetes II. The replacement of Dryton was perhaps be connected with these events. Timokles is given the title *hêgemôn ep'andrôn* and *phrourarchos*, or garrison commander. He is followed by two officers, Aristarchos son of Diognetos and Kriton son of Archas who are given the title *hêgemônes exô taxeôn*. Next in the list comes an individual called Aristomachos son of Aristomachos *exô taxeôn*. Then three sons of officers ([ἡγ]εμόνων υἱοὶ), seemingly all brothers, called Dionysios Panteles and Diogenes sons of one Dionysios, are mentioned. Then finally, at the very bottom of the group comes Eudoxos son of Timokles, who may well be the brother of the unit commander and who is given the title of *grammateus* (secretary) of the *syntaxis* (γραμματεὺς συντάξεως). We shall examine these various rank titles in turn.

The title given to the commander of this unit is *hêgemôn ep'andrôn*. The precise meaning of this term is obscure and has been much debated.[79] It could be

---

[77] Cf. Launey p. 41 n. 7 on the *politikoi*.

[78] Cf. Jan Krzysztof Winnicki, *Ptolemäerarmee in Thebais* (Wroclaw 1978) p. 13.

[79] Cf. eg. Lesquier pp. 84-7.

argued that the term was restricted to regimental commanders, perhaps even commanders of infantry regiments.[80] Feyel has argued that the term was not a specific rank in between that of *stratêgos* and the *hêgemones* of a unit, but rather has a more general sense and was given to the commander of a detachment.[81] The title seemingly pre-dates the reforms of the 160's, although the chronology of some of the inscriptions and papyri mentioning the rank is highly contentious, and so it would be well not to be too dogmatic on the matter. Mitford gathered the occurrences of the title in Cypriote inscriptions, some of which appear to date to the second half of the third century.[82] Further work on the dating of the Cypriote material in the future might result in the lowering of the dates of this material so as to allow it all to be accommodated in the period after the infantry reform of the 160's.

Next in the Lefebvre Stele come two officers, Aristarchos son of Diognetos and Kriton son of Archas who are given the title *hêgemones exô taxeôn*. In *Prosopographica Ptolemaica* the earliest document mentioning the rank of ἡγεμὼν τῶν ἔξω τάξεων has a date of 18 August 134, other than one inscription which is listed with a date earlier than the 160's.[83] This inscription (Figure 5) mentions an officer, a *hêgemôn tôn exô taxeôn*, a Theran called Apollonios son of Sosibios (*PP* 2105). In my view the traditional date of this inscription has been established by a misunderstanding, and it needs to be down-dated to after the infantry reforms of the 160s. Apollonios left the following dedication at to the Samothracian deities known as the Kabeiroi at Koptos.

Θεοῖς μεγάλοις Σαμοθρᾳξι
'Απολλώνιου Σωσιβίου
Θηραῖος, ἡγεμὼν τῶν
ἔξω τάξεων, σωθεὶς
ἐκ μεγάλων κινδύνων, ἐκ-
πλεύσας ἐκ τῆς Ἐρυθρᾶς
θαλάσσης εὐχήν.

To the great Samothracian gods
Apollonios son of Sosibios
a Theran, a *hêgemôn tôn*
exô taxeôn, being saved
from great dangers sailing
out of the Red
Sea (erected this in fulfilment of) a vow.

---

[80] Cf. Holleaux, *Études* III p. 3.

[81] Feyel, *Rev. arch.* 1935, 2 pp. 49-50 'est la fonction de commendant d'armes ou de chef de détachement'.

[82] *Opusc. Athen.* 1 (1953) p. 153 n. 73.

[83] *PP* 2109.

Figure 5. *OGIS* 69. Dedication of Apollonios son of Sosibios a Theraian.

When Dittenberger published this inscription he noted that Egypt exercised firm control of the island of Thera during the reign of Ptolemy III Euergetes I (246-222), however he did not rule out a later date.[84] When Breccia re-published this inscription the date had become 'Probabilmente regno di Tolemeo III',[85] and it has remained 'prob. 246/21' in *PP*. In fact there is no need to date this inscription to the reign of Ptolemy III. A Theran could have been in Ptolemaic service as a mercenary at any time during the existence of the Ptolemaic state. The letter forms (see Figure 5) point to a date in the second century, in particular the broken-barred alpha and the theta with a stroke. A date after the infantry reform of the 160's seems quite suitable for this inscription and it is therefore possible that the rank of *hêgemôn tôn exô taxeôn* was introduced as part of the reforms of the 160's. It must be conceded, however, that the occurrences of this term are too sparse and uncertain

---

[84] *OGIS* 69 p. 122 n.2.

[85] Evaristo Breccia, *Catalogue Général des Antiquités égyptiennes du Musée d'Alexandrie - Iscrizioni Greche e Latine. Vol 1 Nos 1-568* (Cairo 1911) no. 109.

in date to allow any firm dating of its introduction. The suggestion advanced above may have to be rejected in light of future research.

As was the case with the terms *hêgemones ep'andrôn* and *hoi exô taxeôn* discussed above, a number of explanations for the term *hêgemones exô taxeôn* have been advanced previously, but none of them are completely convincing.[86] The unit called 'Those formerly with Dryton' listed in the Lefebvre Stele has two such officers in its headquarters. Presumably two of the *exô taxeôn* of the six *sêmaiai* reported to them. In my view the heralds may have reported to one and the standard-bearers to the other. We do not know the precise way in which the tasks of administering, feeding and equipping the *sêmaia* were divided between the herald and standard-bearer, but it would be reasonable to assume that the two *hêgemones exô taxeôn* divided the regimental supply system along the same lines.

To summarize, the *hêgemôn ep'andrôn* commanded the fighting personnel of the maniples, the officers and soldiers, while the two *hêgemones exô taxeôn* commanded the 'under-officers' supporting the *sêmaia*. For this reason they were given these two different titles.

The fourth person listed in the regimental headquarters, Aristomachos son of Aristomachos, is listed as an *exô taxeôn* but not an officer (*hêgemôn*).[87] He is therefore to be regarded as an under-officer serving at regimental and not at manipular level. I believe it is probable that Aristomachos held the rank of *hypêretês*, or 'attendant', although the stele simply lists him as an under-officer. One notes that the rank of *hypêretês* is not listed in the Lefebvre Stele at all, at either manipular or unit level. The rank is, however, fully attested in the papyri for this period. It also appears in the later Roeder Stele from Hermopolis, discussed below in the following section, where a single *hypêretês* is listed among the ranks of the sixth and last maniple of the regiment. This seemingly proves that the *hypêretês* was an 'under-officer' belonging to the *exô taxeôn,* but that the rank was only found at regimental level. I believe that Aristomachos son of Aristomachos held the rank of *hypêretês* even though this is not expressly stated in the inscription.

The rank of *hypêretês* (*PP* 2435-2452) is attested in the Ptolemaic army both before and after the army reform of the 160's. This is hardly surprising as *hypêretês* is a rank title known to have existed in Greek armies from Classical times onwards.

---

[86] Holleaux, *Études* III p. 11 argued that the term signified officers without effective command, cf. Van't Dack, *Ptoemaica Selecta* pp. 70-71.

[87] According to Holleaux, *E´tudes* III p. 2 n. 2 the omission of the word *hêgemon* is due to a mistake on the part of the sculptor of the inscription. Thus Aristomachos son of Aristomachos should be regarded as a third *hêgemon exô taxeôn* (an argument which is accepted by Étienne Bernand, *Inscriptions grecques d'Hermoupolis Magna et de se nécropole* (Cairo 1999) p. 31). I consider this hypothesis to be most unlikely. It is hardly probable that the sculptor would carve the heading *hêgemones exô taxeôn*, list two officers then carve the heading again (making a mistake) and then list a third officer. Elsewhere his practice is to list all individuals under one heading. It seems clear that the heading *exô taxeôn* is deliberate, and is put there because Aristomachos son of Aristomachos is not an officer.

It was first used of the baggage-carrier who accompanied each hoplite, but later came to be applied to a trusted common soldier placed in charge of the baggage-carriers.[88] In Alexander's army soldiers selected from the ranks during the suppression of the Opis mutiny were handed over to the *hypêrêtai* for punishment, so by this time the term seems to be used as a rank title for one grade of 'under-officer' between the common soldier and the officer proper.[89]

One assumes that the role which the *hypêretês* played in Hellenistic armies in general and the Ptolemaic army in particular became more closely defined. We lack clear comparative material from other Hellenistic armies however, and Roman comparative materials are not very helpful in this case. Polybius mentions the rank once during his description of the battle of Ilipa, but it is difficult to establish whether he is using the term in a specific way, or as a general term for an 'under-officer'. He (11. 22. 4) mentions that as soon as it was light, Publius Scipio sent a message to the tribunes of the army via his *hypêretai*. At first glance it would seem that these particular *hypêretai* correspond to the *beneficiarii* who served as orderlies to senior officers of the legion in Imperial times.[90] In a parallel passage Livy (28. 14. 7) tells us that the same message had been conveyed round the army the previous day by *tessera*.[91] It seems that Polybius may therefore be referring to the *tesserarii* of the army in this passage, and is using the term *hypêretai* in this passage as a standard Greek term for 'under-officers'.

Elsewhere in the Lefebvre Stele, in Column I line 61, one Sopatros son of Kassiodoros is listed as an *archypêretês xenikou*, or 'head attendant of the mercenary force'. He presumably performed the same function as the *hypêretês* at a level above the unit. It may be possible to define the duties of the *hypêretês* by reference to what we know of the tasks of the *archypêretês*. In the course of his examination of the inscription from Amphipolis listing military regulations for the Antigonid army, Feyel analyzed the role of the *archypêretês* in the Ptolemaic papyri.[92] In the first document Feyel studied, an ostrakon from Thebes dating to the second century,[93] an *archypêretês* writes to two individuals of unknown status concerning the dispatch of pay to the *hekatontarch* Menon.

'Απολλών(ιος) ἀρχυπη(ρέτης) Θεοδώ(ρωι) καὶ Δημω(...)
χαί(ρειν). Δοθήτω Μένονι .. ρ́
τὸ καθῆ(κον) μέτρη(μα) καὶ ὀψώνιον.

---

[88] Cf. Thuc. 3. 13. 3; Xen., *Cyr.* 2. 4. 4, 6. 2. 13, *Hipparch.* 4. 4

[89] Diod. 17. 108. 2.

[90] Graham Webster, *The Roman Imperial Army of the First and Second Centuries A.D.* (second edition, London 1979) p. 121.

[91] Cf. Walbank, *Commentary* II p. 299 on the problems surrounding this disparity.

[92] Feyel, *Rev. arch.* 1935, 2 pp. 42-5.

[93] Ulrich Wilken, *Griechische Ostraka aus Aegypten und Nubien* II (Leipzig, Berlin 1899) p. 406 no. 1538.

Apollonios the *archypêretês* to Theodoros and Demo(...)
hail! Give to Menon the Hekatontarchos
the proper ration allowance and pay.

The use of the palaeographical symbol ρ́ to indicate that Menon holds the rank of *hekatontarchos* clearly dates this ostrakon to after the infantry reform of the 160s.[94] Of the two words *metrêma* and *opsônion* the former means 'rations' and the second 'pay'. The second element was always paid in cash, but the first could be paid either in cash or in kind.[95] The two elements are nevertheless always listed separately.[96] It seems then that the *archypêretai* and *hypêretai* were concerned with the disbursement of pay and allowances, the former at formation level and the latter at regimental level. Feyel suggested that Theodoros and Demo(...) may have been *hypêretai*, bankers or cashiers, at any rate they would have been concerned with financial affairs. In the case of the individuals listed on the Lefebvre Stele, Sopatros son of Kassiodorus was probably responsible for disbursing pay to the whole mercenary garrison at Hermopolis Magna, which he did through the *hypêretai* of the different units stationed there. This is the pattern suggested by Lesquier, that an *archypêretês* distributed pay and rations at formation level through the *hypêretai* of the various corps and units in the formation or garrison.[97]

We do not know if the three 'sons of officers', who are listed next in the Lefebvre Stele, were attached to the regimental headquarters in a military sense or if they are simply listed at here for convenience. If the title does have a real military significance, it may be that they really are sons of officers attached to a regimental headquarters for the purposes of training. That is, they may be considered the equivalents of 'officer cadets' or 'ensigns'.

Finally Eudoxos son of Timokles the *grammateus*, or secretary, of the unit is listed as the last member of the headquarters element of 'Those formerly with Dryton'. We do not know what the duties of the *grammateus* (or secretary) of the unit were. They are presumably separate from the provision of pay and rations,

---

94 Wilcken (loc. cit.) dates the document to 'Aus dem II. Jahrh. v. Chr.'

95 G. T. Griffith, *The Mercenaries of the Hellenistic World* (Cambridge 1935) p. 279; Launey, *Recherches* II p. 735.

96 Walbank, *Commentary* II p. 161.

97 Lesquier p. 101. According to Lesquier this distribution was done upon the orders of the *grammeteus*, but it seems that the *grammateis* operated along a different chain of command from the *hypêretai*. In contrast Feyel (*Rev. arch.* 1935, 2 p. 61) suggested that in the Macedonian army under Philip V the *hypêretês* was found at sub-unit level (the σπεῖρα), the *archypêretês* at unit level (the chiliarchy) and the *grammateus* was found at the level of the army (*statêgia*). This may have been true of the Antigonid army at an earlier period, but the evidence cited above suggests that in the Ptolemaic army after the reform of the 160s the *hypêretês* and *archypêretês* are found at higher levels in the chain of organization, and that the *grammateus*, who is also found at unit level, had different responsibilities. It is possible that the same pattern of responsibilities was also to be found in the army of Philip V.

which were the duties of the *hypêretês*. It would be a reasonable assumption that the *grammateus* was responsible for all the administrative paperwork of the unit.

Clearly the officers of the *sêmaia* would report directly to the *hêgemôn ep' andrôn* of the unit for their orders, and it is reasonable to assume that the three *exô taxeôn* of the *sêmaia* were responsible for different tasks in the administration of the *sêmaia*, and would report to different individuals at unit level. But it is not clear exactly who reported to whom. If we consider the *sêmeia* to be the equivalent of the modern infantry company, the nearest equivalent to the rank of *ouragos* would be that of Company Sergeant-Major. In battle he would be responsible for ensuring no-one left the ranks, and we may assume that his principal duty in time of peace was to make a daily note of who was present in the *sêmaia* and who was absent. He presumably reported to the *grammateus* of the unit. This may be the reason why, just as the *ouragoi* are listed at the bottom of each *hekatontarchia*, the *grammateus* is listed at the very bottom of the headquarters element, even though he is obviously more important in status than the sons of officers. Indeed the *grammateus* is likely to have been the most important officer in the headquarters after the commander. In this case the two are brothers.

The herald and the standard-bearer might perhaps be compared to Staff-Sergeants in a modern army. How they divided the other administrative duties of the *sêmaia* between them is unknown. In the Roman army the standard-bearer looked after the money deposited by men in the century or maniple.[98] It is possible that the standard-bearers in the Ptolemaic army also had some financial responsibilities. The ostrakon from Thebes discussed above indicates, however, that pay and rations were disbursed through the *hekatontarchoi*.

This concludes our discussion of the members of the headquarters element of a Ptolemaic infantry regiment. We must now attempt to establish by what name the new six-maniple unit in the Ptolemaic army was known.

The title given to Eudoxos son of Timokles in the Lefebvre Stele is γραμματεὺς συντάξεως, so it would be reasonable to suppose that the name given to the new unit of the army was *syntaxis*,[99] and that the correct expanded title of the unit in question is 'the *syntaxis* formerly with Dryton'. The term also occurs in *PFayum*, 302, a papyrus dated to the second century BC, recording an application from certain *machairophoroi* (literally 'sword bearers' but in the Ptolemaic army probably some form of attendants such as baggage-carriers). The *machairophoroi* ask if it seems that another *syntaxis* has been given the ration allowances and pay for five men for the months of Thot and Phaophi which were intended for them.[100] Here, then, is a second secure reading of the term *syntaxis*. From the

---

[98] G. R. Watson, *The Roman Soldier* (London 1969) pp. 51-2.

[99] Cf. Van't Dack, *Ptolemaica Selecta* p. 71 n. 22.

[100] ἐ]ὰν φαίνηται σύνταξιν δοθῆναι ἡμῖν τὸ μέτρημα καὶ τὸ ὀψώνιον τοῦ Θωὺθ καὶ Φαῶφι ἀνδρῶν πέντε.

context it would be difficult to interpret this term as anything other than the title for a unit. It must be admitted that these two readings of the term are the only secure ones in Ptolemaic documents.[101]

Recalling that the smallest *pentekontarchia* recorded in the Lefebvre Stele numbers 31 men plus one officer, the minimum establishment strength of the new infantry unit would have been 744 men plus 59 officers and 'under-officers'. The unit would be close to a thousand men strong and the title of chiliarchy might have been quite appropriate. We have already noted, however, that none of the occurrences of the title *chiliarchos* listed in *Prosopographica Ptolemaica* have a date later than the 160's, and there is no evidence for the term chiliarchy being used after that date. Possibly the re-organization of the infantry was carried out on the basis of the previously existing chiliarchies.

Lesquier suggested that the titles *hêgemonia* and *syntagma* may have been used successively for the Ptolemaic regiment.[102] We first turn to examine his arguments for the *syntagma*. The word occurs, seemingly in an abbreviated form, in a number of documents contained in the volume *PPetrie* III. Grammar dictates that the word can not stand as it is, though there are no signs of abbreviation.[103] Lesquier proposed that restoration of the term as a contraction was inevitable, and suggested that it should be restored as συνταγμα(τάρχης), noting that the term *syntagma* appears in Asclepiodotus (2. 8) as the title of a unit of 256 men.[104] Bell later published other examples of the term occurring in abbreviated form, and accepted Lesquier's suggestion.[105] Nevertheless Lesquier's conjecture remains far from certain, and is in any case based on evidence from the third century, before the reform which is the subject of the present study. Zucker went further and proposed that the sub-units listed on the 'Jouget' and 'Roeder' Stelai from Hermopolis Ma-

---

[101]In *P. Grenf.* I 10, 8 the paymaster of a '*[---]taxis* of the infantry' ([---]τάξεως πεζῶν τακτόμισθος) could perhaps be restored to read *[syn]taxis*, though a number of other restorations are possible. This restoration was proposed by Lesquier (and Holleaux, *Études* III p. 5) although he did not consider that the word was given to a unit, but rather to 'a group of troops, a garrision'. This stems from his interpretation of the group of officers listed after Timokles son of Timokles as being a command element of the garrison rather than the command element of 'Those formerly under Dryton', which clearly refers to a unit. However the papyrus is dated to 174 BC and therefore to before the infantry reform. Perhaps the word had come into use as an alternative for the chiliarchy prior to the reform.

[102]J. Lesquier, 'Note sur une unscription d'Ashmounein' *Rev. phil.* 32 (1908) pp. 215-25, especially p. 222.

[103]Cf. Van't Dack, *Ptolemaica Selecta* p. 58.

[104]Lesquier pp. 92, 94-7, 195-7, 345, 366-7.

[105]H. I. Bell, 'Notes on Early Ptolemaic Papyri. 1. An Early Ptolemaic Contract of Loan' in ed. Ulrich Wilcken, *Archiv Für Papyrusforschung und Verwandte Gebeite* 7 (Berlin 1924) pp. 17-27 esp. p. 24. Also accepted by Fritz Uebel, *Die Kleruchen Ägyptens unter den ersten sechs Ptolemäern* (Berlin 1968) p. 378 cf. p. 225 n. 2 who notes that Peremans and Van't Dack (*PP* ii p. xxii) suggested that it might be an abbreviation for a unit not for the unit commander.

gna should be identified as a *syntagmata* of 250 men.[106] Asclepiodotus uses the term *syntagma* for the maniple, to which the 'under-officers' are attached, not for a higher level of organization, as has been noted by Van't Dack.[107] The whole unit at Hermopolis Magna cannot therefore have been a *syntagma* and we know that the maniple was called a *sêmeia*. Asclepiodotus' use of the term *syntagma* for the maniple is puzzling. Though perhaps appearing in slightly different spellings, the term *sêmaia* is used for the Roman maniple by Polybius, and it is also the term used for the maniple in the Ptolemaic Army. Finally it should be noted that although many modern authorities have asserted that the authors of the Greek *Taktika* are describing the army of Alexander the Great in their works, this is patently false and has led to great confusion. There is no evidence whatsoever for the existence of a *syntagma* of 256 men in the army of Alexander.

Lesquier's second suggestion that *hêgemonia* (literally meaning 'a command') may have been used as the title of the regiment has likewise received some support.[108] Lesquier noted a document of the second century which mentions one 'Pasis, son of Petenephotes, a Persian, a *hekatontarchos* of the Akorites of Asklepiades' command'.[109] On the basis of this document Lesquier suggested that the name of the regiment in the Ptolemaic army may have changed from *syntagma* to *hêgemonia* in the second century. In a second papyrus dated to 29 July 127, a *grammateus* (secretary) 'of the phalanx over which Polianthes holds command' (γραμματεὺς φάλαγγας) is mentioned. Later on in the same papyrus the term *hêgemonia* is used of the same formation which is commanded by Polianthes. Bearing in mind that the term *hêgemonia* literally means 'command', and that the term is used interchangeably with the word phalanx in the second papyrus, the most logical conclusion is that the word is used in a general sense, and was not used as the official title for any specific level of command.[110]

It seems that wherever the terms *syntagma* or *hêgemonia* appear in the papyri, it is more likely that both are used in an informal sense, not as official titles of a level of military organization. *Syntaxis* remains the only suitable term for regiment which appears in a certain reading and context in any Ptolemaic document, and therefore should be accepted as the standard term for the Ptolemaic six-maniple regiment until new contradictory evidence comes to light.

---

[106] F. Zucker, *Doppelinschrift spätptolemâischer Zeit aus der Garnison von Hermopolis Magna* (= *Abhandlungen der Preußischen Akademie der Wissenschaften, Jahrgang 1937, Philosophisch-historische Klasse Nr. 6*, Berlin 1938) p. 29 ff. Followed by Van't Dack, *Ptolemaica Selecta* pp. 56-7 and Étienne Bernand, *Inscriptions grecques d'Hermoupolis Magna et de se nécropole* (Cairo 1999) p. 51.

[107] Van'T Dack, *Ptolemaica Selecta* p. 57.

[108] Lesquier pp. 80, 84, 96; Zucker, *op. cit.* p. 30; cf. Van't Dack, *Ptolemaica Selecta* p. 55.

[109] *PReinach* I 26 lines 3-5 Πᾶσις Πετε[ν]εφώτου Πέρσης, | ἑκατόνταρχος τῶ[ν] τῆς 'Ασκληπιάδου | | ἡγεμονίας 'Ακωρ(ι)τῶν.

[110] Cf. Van't Dack, *Ptolemaica Selecta* p. 55 n. 21.

Section 1 f

The Ptolemaic Infantry Regiment in the First Century

Two other inscriptions from Hermopolis Magna, known as the 'Jouget' and 'Roeder' stelai, give us information on the structure of a Ptolemaic infantry regiment at a later date.[111] Both stelai have the same tympanum and start with an introductory text. In the case of the Roeder Stele, where the introduction is better preserved, the stele is given a date of 25 January 78 and we are informed that the two stelai together list a regiment called the 'Apolloniate Mercenaries' (ξένοι Ἀπολλωνιᾶται). The Jouget Stele was carved a little earlier and belongs to the year 80/79 BC.[112] Beneath this common introductory element the Jouget Stele (Figure 6), which was published first,[113] lists the first three *sêmaiai* of the regiment, while the Roeder Stele (Figure 7) lists the remaining three *sêmaiai*. When the listing of a new *sêmaia* starts its number is given to the left of the column (see Figure 6). The stele includes a number of individuals with Semitic personal names as well as ones with Greek or other origins. Zucker, who published stele R in a monograph devoted to a study of both stelai,[114] suggested that the regiment was composed of mercenaries originally from Idumaea, who were now resident in Egypt. If his interpretation is correct it is possible that this Idumaean community had been displaced by the expansion of the Jewish state in the last decades of the second century BC.

Though the basic outline of the Ptolemaic infantry regiment remains the same, the Jouget and Roeder Stelai clearly demonstrate that some organizational changes have taken place. In the following paragraphs these differences will be examined, as well as any further information which the stelai give us as to the organization of the Ptolemaic infantry regiment in general.

---

[111]Étienne Bernand, *Inscriptions grecques d'Hermoupolis Magna et de se nécropole* (Cairo 1999) no. 5 pp. 36-56 (Jouget Stele, bibliography p. 36), no. 6 pp. 57-72 (Roeder Stele, bibliography p. 57).

[112]Étienne Bernand, *Inscriptions grecques d'Hermoupolis Magna et de se nécropole* (Cairo 1999) p. 47.

[113]P. Jouget, 'Inscriptions grecques d'Égypte III. Hermopolis Magna (*Ashmounein*)' *BCH* 20 (1896) pp. 177-191; J.G. Milne, *Catalogue général des antiquités égyptiennes du Musée d'Alexandrie. Greek Inscriptions* (Oxford 1905) pp. 25-7 no. 9296; *SB* 4206.

[114]Friedrich Zucker, *op. cit.* Improved readings and further fragments of stele R were subsequently published in Friedrich Zucker, 'Nachträge zur 'Doppelinscrift spätptolemäischer Zeit aus der Garnison von Hermopolis Magna' *Aegyptus* 18 (1938) pp. 279-284. Zucker argued that the mercenaries were originally from the Idumaean city of Apollonia, but Rapaport argued they were from Marissa (Étienne Bernand, *Inscriptions grecques d'Hermoupolis Magna et de se nécropole* (Cairo 1999) p. 48).

Figure 6. The 'Jouget Stele' from Hermopolis (photo: P. M. Fraser).

Figure 7. The 'Roeder Stele' from Hermopolis (after Zucker, *op. cit.*).

The regimental commander of the 'Apolloniate Mercenaries' is one Herakleides son of Apollonios, one of the 'First Friends', *hêgemôn*, and *phrourarchos* (garrison-commander). He is listed as being in personal command of the fifth *sêmeia*. The reason for this is, presumably, that when the regiment was drawn up in *acies triplex* the fifth *sêmeia* would be drawn up at the back on the right-hand side, and when the regiment was drawn up in *acies duplex* it would be stationed in the centre of the rear line. In was normal for commanders to position themselves either at the right or centre of the unit. Despite damage to the surface of the stone, and so to the text, it is clear that all the other *sêmeiai* are commanded by an officer given the simple title of *hêgemôn*, they all have an *ouragos* and standard-bearer, and four pentekontarchs, but the rank of hekatontarch is gone.[115]

Although we have already noticed that the complement of officers in the *sêmeiai* listed in the earlier Lefebvre Stele is irregular, the *hekatontarch* is present. In another group of documents, namely in the dossier of papyri concerning the Egyptian *sêmeia* which took part in the Palestinian war of 103-101 BC,[116] although the correspondence is sometimes addressed to Pates alone, who is presumably the senior hekatontarch of the *sêmeia*, more usually the correspondence is addressed to Pates and Pachrates the *hêgemones* of the *sêmeia*. From this fact it would be reasonable to assume that it remained normal practice to have the *sêmeia* commanded by two hekatontarchs down to this date. Other than these two, the latest certain reference to a hekatontarch seems to be to one Pasion, a hekatontarch in Askepiades' *hêgemonia* of the soldiers of Akoris in documents of 103 BC (*PP* 2325). Unless the regiment listed in the Jouget and Roeder Stelai is exceptional, it seems that the title of hekatontarch fell out of use in the Ptolemaic army shortly after.

The previous establishment of one herald per *sêmaia* is also reduced to one per regiment, for only the fourth *sêmaia* includes a ταγματι(κὸς) κῆ(ρυξ) who also happens to be a high-priest.[117] The sixth *sêmeia* also had a *hypêretês*, listed after the pentekontarchs, who presumably acted as the *hypêretês* for the whole regiment. The second *sêmeia* only has three pentekontarchs, but it also has an officer who holds a title, given in abbreviated form on the stele, beginning with the letter *gamma* (see Figure 8). He may be the *grammateus* of the regiment, doubling up as one of the pentekontarchs of the second *sêmeia*. The second *sêmeia* included a sacred-flautist and the fourth a sacred-psalmist, but these are titles connected

[115]Cf. Étienne Bernand, *Inscriptions grecques d'Hermoupolis Magna et de se nécropole* (Cairo 1999) p. 66.

[116]E. Van't Dack, W. Clarysse, G. Cohen, J. Quaegebeur & J. K. Winnicki, *The Judean-Syrian-Egyptian Conflict of 103-101 B.C.. A Multilingual Dossier Concerning a "War of Sceptres"* (*Collectanea Hellenistica I*, Louvain 1989) pp. 37-81.

[117]*Aegyptus* 18 (1938) p. 282; cf. Jan Krzysztof Winnicki, *Ptolemäerarmee in Thebais* (Wroclaw 1978) p. 15 who also noted the contrast with Stele L, and Étienne Bernand, *Inscriptions grecques d'Hermoupolis Magna et de se nécropole* (Cairo 1999) p. 68.

Figure 8. Detail from the 'Jouget Stele' giving the officers of the second *sêmeia* of 'The Apolloniate Mercenaries'. The symbol defining the rank of the person named in the sixth line down, the letter *gamma*, cannot be interpreted with any certainty, but it could possibly stand for *grammateus* (photo: P. M. Fraser).

with the distinctive religious practices of this Idumaean unit of religious exiles, and have no military significance.[118]

In his initial publication Zucker (p. 28) considered that the stelai were substantially complete, though damaged at the bottom, and calculated that in addition to the officers and staff, the first *sêmeia* had a strength of 96 rankers, the second 55, the third 68, the fourth 64, the fifth 62 and the sixth 61. These figures have to be treated with some caution as the fragments found subsequently to Zucker's initial publication of Stele R add at least 13 more individuals to the total for the fourth *sêmeia* and probably at least 8 to the fifth.[119] Nevertheless it seems certain that this regiment was significantly under-strength. It is uncertain whether the first *sêmeia* was maintained at a greater strength than the others deliberately, or whether this was simply due to chance. Below the third *sêmeia* are listed some seventeen or so native Egyptian 'Royal Sword bearers selected from the Companies' ('Εγλελοχισμ-

[118]Thus Zucker, *op. cit.* p. 32 contra Lesquier p. 97.

[119]*Aegyptus* 18 (1938) p. 281.

ένοι μαχαιροφόροι βασιλικοί). We do not know what role these troops performed. Most probably they are gendarmes or baggage-carriers attached to the regiment.[120]

In the first century, therefore, the command and administrative staff of the Ptolemaic infantry regiment had been somewhat slimmed down. The breaking up of the Regimental Headquarters and its distribution among the *sêmeiai* may just be an expedient practised in this regiment which was considerably below strength. The reduction of the numbers of military heralds from one per *sêmeia* to one per regiment, and more importantly, the discontinuation of the rank of hekatontarch, more probably reflect changes which had taken place throughout the Ptolemaic army as a whole. It is possible to suggest who may have been responsible for introducing this second reform in the organization of the Ptolemaic infantry regiment.

A prominent figure in Ptolemaic military circles during these years was one Philostephanos, a possible descendant of the famous third-century Greek writer Philostephanos of Cyrene. Philostephanos commanded the army of Ptolemy IX Soter II Lathyros against the Jewish king Alexander Jannaeus in 103 BC. His greatest victory was achieved at the battle fought at Asophon, on the east bank of the Jordan, where, commanding an army of only 30,000 foot and horse, he inflicted a defeat on Alexander's army of 50 or 80,000 troops by a skilful manoeuvre on the battlefield. Josephus (*AJ* 13. 340)[121] describes Philostephanos as a military writer (ὁ τακτικός), and Plutarch has preserved one of his fragments in the *Life of Lycurgus* (23. 1). Ptolemy Lathyrus, then an exile, only re-conquered Egypt in 88-9 BC and ruled until 81. The reform may have taken place during this period.

Section 1 g

The Author of the Reform

Our discussion of the composition of the Ptolemaic infantry maniple and regiment are concluded. We shall now turn to the question of who might have been the author of the reforms of the 160s, which occurred during the reign of Ptolemy VI Philometor. According to Polybius (39.7) Philometor was 'a gentle and good king'. He never put to death any of his Friends upon any accusation whatsoever; nor did any of the Alexandrians suffer death by his will. Polybius also tells us that despite

[120]Cf Jan Krzysztof Winnicki, *op. cit.* p. 16. Van't Dack, *Ptolemaica Selecta* p. 60 thinks they may be an attached *lochos*.

[121]I would like to thank Richard Taylor for initially bringing my attention to this passage.

his sensitive nature, Philometor showed courage and presence of mind in political crises and on the battlefield. He was extremely popular with his troops: a decree set up by his Cretan auxiliaries at Delos calls him 'scrupulous, pious and of all men the most humane ... showing a great spirit in all his dealings'.[122] In Rome Philometor found an ardent supporter in Cato the Censor, who spoke against Thermus on his behalf.[123] Cato may have met Philometor during the king's visit to Rome in 164, and would have been impressed by the young monarch's character. The overall picture we have of Philometor in the sources suggests a capable monarch, able to institute reform in the kingdom.[124] We have seen, however, that the infantry was reformed shortly before the year 163/2. Philometor was born in 186, or more probably in 184/3,[125] therefore he was probably only a teenager at the time of the reform. It is most unlikely that Philometor himself was directly responsible for the reform.

The 160's were a decade of turbulent events, involving the disastrous Sixth Syrian Year, native revolts and the struggle for power with his brother Ptolemy Euergetes, during which the leading figures of the administration probably changed considerably. It is difficult to find any one individual who might be credited with the reform. Kallikles son of Kallikles of Alexandria, however, stands out among his contemporaries as the most likely candidate. Kallikles is only attested in a couple of honorific inscriptions from Cyprus which list his military titles. One of these, unique in Ptolemaic court titulature,[126] is that of 'instructor in tactics to the king'. Both inscriptions come from Paphos, and are of great interest. They are given below with the restorations of Mitford, who dates them both to 154.[127]

[Βασιλέα Πτολεμαῖον Θεὸν]
[Φιλ]ομήτο[ρα] Κ̣[αλλικλῆς]
[Κ]αλλικλέ[ο]υς 'Αλε[ξανδρεὺς]

[122] *Insc. de. Délos*, 1518, lines 6-7; Walbank (*Commentary* III p. 738) suggests that Polybius may have formed his favourable opinion of Philometor in part through his friendship with Menyllus of Alabanda. This is unnecessary, however, as it is reasonably certain that Polybius would have met Philometor personally during the visit of the latter to Rome in 165.

[123] Cf. Alan E. Astin, *Cato the Censor* (Oxford 1978) p. 270.

[124] On Polybius' view of Ptolemy VI see Arthur M. Eckstein, *Moral Vision in the Histories of Polybius* (Berkeley 1995) p. 275 'Even the glorious death of Ptolemy VI in battle does not absolve him completely of Polybian suspicion concerning his essential laziness'.

[125] Volkman, 'Ptolemaios (24)' *PW* xxiii, 2 (1959) 1702-3; cf. Walbank, *Commentary* III p. 312.

[126] Leon Mooren, *The Aulic Titulature in Ptolemaic Egypt, Introduction and Prosopography* (Brussels 1975) p. 21, who notes that whatever his activities in Cyprus, his titles 'must be related to his palace appointment at Alexandria, and it is only to the latter that his court rank can possibly be linked.'

[127] T.B. Mitford, 'The Hellenistic Inscriptions of Old Paphos' *BSA* 56 (1961) pp. 20-22 nos. 53-4.

ὁ ἀρχισωματοφύλ[αξ καὶ ἐπὶ]
τῶν ἐν 'Αλεξανδρε[ίαι ἱππέων]
καὶ διδάσκαλος [τοῦ βασιλέως]
τακτικῶν εὐερ[γεσίας ἕνεκα]
τῆς εἰς ἑαυτὸν καὶ [τὰ τέκνα ?]

[Καλλι]κλῆν Καλλικλέο[υς τῶν σωματοφυλάκων Πτο]
[λ]εμαίου ἰλάρ[χην, τῶν ἐν 'Αλεξανδρείαι τεταγμένων]
ἱππέων καὶ εὐωνύμων [ἐπιστττά]ṭῃν καὶ διδάσ[καλον]
τοῦ βασιλέως τῶν τ[ακτι]ḳῶν *vacat* ἡ συναρχία τῶν ἐν Πάφωι.
στρατηγῶν καὶ ἐστρα[τηγ]ηκότων ἀρετῆς ἕνεκα καὶ
εὐνοίας τῆς εἰς Βασιλέ[α] Πτολεμαῖον καὶ Βασίλισσαν
Κλεοπάτραν τὴν ἀδελφήν, Θεοὺς Φιλομήτορας, καὶ τὰ τέκνα
καὶ τῆς εἰς ἑαυ[τ]ὴν εὐεργεσίας καὶ πρὸς ἅπαντας δικαιοσύνης.

The differences in titulature between the two inscriptions are striking. It has been noted that not all the restorations are inevitable,[128] but indeed it is hard to see how the restorations of either inscription could be improved upon other than in detail. Mitford believed the differences to be due to the fact that Kallikles himself dictated the first inscription, a dedication on his part to Philometor, using the correct aulic ('courtly') titulature, while the second inscription was a dedication to Kallikles by the past and present *stratêgoi* of Paphos 'who in their remoteness from the Ptolemaic ruling class find it necessary to interpret his rank and offices'. This would be a most strange phenomenon.

The precise nature of the activities of Kallikles in Cyprus is unknown.[129] He may well have been sent to the island as a military advisor to reform the garrison of this key Ptolemaic possession. Mitford observed that under the reign of Philometor the organization of the garrison changed. The casual mercenary structure of the garrison was altered, and specialist technical units are attested for the first time.[130] The military 'architects stationed on the island' under the command of one Karpion are mentioned in the inscriptions of the island.[131] A regiment (*tagma*) of veteran

[128]Van't Dack, *Ptolemaica Selecta* p. 61.

[129]Cf. Leon Mooren, *The Aulic Titulature in Ptolemaic Egypt, Introduction and Prosopography* (Brussels 1975) p. 21 'Whatever his activities there, they must be related to his palace appointment at Alexandria, and it is only to the latter that his court rank can possibly be linked'.

[130]T.B. Mitford, *Opuscula Atheniensia* 1 (1953) p. 150 'the impression is created that Philometor did away with the casual mercenary organization which had satisfied his predecessors, replacing it with a loyal and homogenous force, in which professional ability in the technicalities of war found a proper encouragement'.

[131]*BSA* 56 (1961) p. 23 no. 58.

artillerists and a regiment of junior artillerists (τάγμα τῶν πρεσβυτέρων (or νεωτέρων) ἀφετῶν) honour a citizen of Patara in Lycia, who also seems to be a military architect.[132] It is possible that the soldiers of these two regiments were Lycians too. After the death of Philometor and the accession of Euergetes the island's garrison was concentrated at the capital Paphos, at Salamis and also possibly at Kition. Cilician and Lycian regiments were stationed at Paphos and a Cretan regiment at Salamis.[133] Ionian and Achaean regiments are also known. The regiment of Achaeans may well have been composed of fugitives from the Roman conquest of the Achaean League.[134]

Kallikles holds the post of *archisomatophylax* or 'head bodyguard', which might be translated as 'senior staff officer' in modern parlance, but is really an honorific court title . He is also named as a squadron commander (*ilarchês*) of the cavalry in Alexandria. By this Mitford thought the palace cavalry (περὶ τὴν αὐλὴν ἱππεῖς) was meant. It may also be a purely honorific title, or Kallikles may have been an eponymous officer in this regiment, though not actually present with his regiment at the time, being seconded for temporary duty in Cyprus.

The other two military titles he holds are quite extraordinary. He is called 'instructor in tactics to the king' (διδάσκαλος τοῦ βασιλέως τῶν τακτικῶν). In this context, of course, tactics means the art of drawing up an army. To what degree this title is honorific, and to what degree Kallikles was actually responsible for instructing the king in tactics is impossible to determine. One would like to believe that, above all else, Kallikles was charged with instructing in the art of drawing up the army in the Roman manner. He may well have had a hand in the army reform of the 160s. The final title Kallikles is given is commander 'of the troops of the left'. The possible implications of this title will be discussed in the next section.

Kallikles is not otherwise known, and we have no knowledge of where he may have acquired his specialist knowledge. He may have visited Rome personally as an ambassador, and it is possible that he may have served as a mercenary officer in the western Mediterranean. In this way he may have acquired direct experience of the Roman military system. Otherwise he may have obtained the information he required to implement the reforms by interrogating the considerable number of Italian mercenaries serving in the Ptolemaic army at the time.

The first Egyptian embassy to Rome was sent in 273, and Romans and other Italians are found in Ptolemaic service from the middle of the third century onwards. The earliest attested of these is one Dinnos, a Roman, a *dimoiritês* in the regiment commanded by Automedon in 252/1 BC.[135] A *dimoiritês* was the sec-

---

[132]W.H. Buckler, *JHS* 55 (1935) pp. 75-8; Launey II p. 1015.

[133]*BSA* 56 (1961) p. 27-9.

[134]*Opuscula Atheniensia* 1 (1953) p. 152 n. 70.

[135]H. I. Bell, 'Notes on Early Ptolemaic Papyri. 1. An Early Ptolemaic Contract of Loan' in ed.

ond-in-command of an infantry file. Among all the Hellenistic armies Romans are found almost exclusively in Ptolemaic service.[136] This could perhaps indicate a right negotiated by an early Ptolemaic embassy to recruit Roman citizens. Some of these Italian or Roman citizens reached quite high rank. The Roman Lucius son of Gaios was commander of the Ptolemaic garrison stationed at Itanos in Crete during the reign of Philopator (217-209 BC).[137] Reinach dated the dedication which Lucius made to Ptolemy Philopator and Queen Arsinoe to the years 216-206, as the king did not marry until after the battle of Raphia, and after 206 the son is also mentioned in official documents. The presence of Lucius is particularly remarkable as at that time Rome was involved in the Second Punic War. Reinach compared the case of Lucius to that of Decius Magius attested in Livy 23. 7-10. Decius Magius was a Capuan who had violently opposed the alliance made between that city and Hannibal in 216. He was subsequently arrested and handed over to the Carthaginians. The ship in which he was being transported to Carthage was carried to Cyrene by a storm however, and Decius Magius was taken before the king in Alexandria, where he stayed, since it was unsafe to return to Capua, and entry into Roman service would make him a traitor to his native city. Reinach suggests that Lucius found asylum in Alexandria on condition of not returning to Roman service. Perhaps he was an officer in a Roman garrison which had capitulated under terms. Reinach further suggests that Lucius may have returned to Rome in 210 when a Roman embassy renewed the treaty (Livy 27. 4; Polyb. 9. 45).[138] Thus even in Rome's greatest times of need, some of her manpower could be found in Ptolemaic service. Lucius must surely have been an experienced officer to have been appointed commander of one of the key Ptolemaic garrisons. Such individuals would have spread knowledge of Roman military systems within the Ptolemaic military and political establishment and would have prepared the way for the reform of the 160's.

Ulrich Wilcken, *Archiv Für Papyrusforschung und Verwandte Gebeite* 7 (Berlin 1924) pp. 17-27 esp. pp. 20, 26; Launey I p. 606 n. 3. To Launey's list now add Νούιος Νουΐου in line 112 (= col. II line 40) of the Lefebvre Stele (Étienne Bernand, *Inscriptions grecques d'Hermoupolis Magna et de se nécropole* (Cairo 1999) p. 34).

[136]Launey I pp. 606-9.

[137]*ICret.* III, iv, 18 (p. 115-6).

[138]A.J. Reinach, 'Incriptions de Itanos' *RÉG* 24 (1911) pp. 400-415 no. III.

Section 1 h

Evidence Concerning Higher Formations

We have already mentioned the relevance of the Greek writers of *Taktika* to discussion of late Hellenistic military practice, with relevance to the *ektaktoi* of the Ptolemaic maniple. We now return to this source to see if any light can be shed on how the Ptolemaic army was organized at formation level: that is above the level of the unit or regiment. According to the scheme presented by Asclepiodotus the maniple of 256 men keeps doubling until a total strength of 16,384 is reached for the phalanx of heavy infantry. This figure is also given by Aelian and Arrian, and so presumably goes back to Poseidonius. Modern commentators have ridiculed this figure. For example in the Loeb edition of Asclepiodotus (p. 257 n. 2) Oldfather states.

> 'The number 16,384 represents, of course, only an ideal for tactical convenience and exactness in manoeuvres. As an ideal or standard theoretical number it does no harm, since no one would dream of allowing it to interfere with practical considerations'.

While not seeking to deny that 16,384 represents an ideal establishment strength rather than the parade state of an 'Asclepiodotan' army, we must ask why do the authors of the *Taktika* stop precisely at that figure rather than, say, double it further?

Under normal circumstances the Roman army was organized into four urban legions (Polyb. 6. 19. 7) numbered I to IV, each with a strength of 4,200 (Polyb. 6. 20. 8), a total of 16,800. During the Second Punic War there were upwards of 20 legions in the field, and in the period which followed there always tended to be a certain number of legions embodied and deployed on campaign or garrison duty at any one time. Nevertheless, the practice tended to be that when a new army was mobilized it normally consisted of four legions, divided between the two consuls. We can see this process in action in 171 when war was declared against Macedon. Livy (42. 31. 2) tells us that four legions were raised, two to be allotted to the consul who would carry out the war against Macedon, and the other two to remain in Italy with the other consul. The legions destined for Macedon were to be levied up to a strength of 6,000 while the ones which remained in Italy 'in accordance with the old custom' were to be levied up to a strength of 5,200 infantry. A total strength of about 16,000 was not the norm for the Roman army alone during the second century. The figure of 16,000 is given as the strength of the Seleucid phalanx at the battle of Magnesia: it is also, incidentally, the figure which Livy gives to the Macedonian phalanx at Pydna.[139] Thus the choice of the number of 16,384 as the strength of the heavy infantry force of the 'ideal' army of Asclepiodotus is quite in line with late Hellenistic practice.

Moving down through the chain of organization Asclepiodotus (2. 10) has the heavy infantry, which he calls the 'phalanx', divide into *diphalangiai* or 'wings' commanded by a *kêrarchês* or 'wing-commander'. The wing further divides into two formations which according to Asclepiodotus are 'even now' called *phalangarchiai* or 'half-wings', commanded in former times by a *stratêgos* but now by a *phalangarchês*, and numbering 4,096.

In support of the organizational framework proposed by Asclepiodotus, we should note that in descriptions of Hellenistic battles the infantry of the centre is frequently divided into two halves, each commanded separately. At the battle of Kynoskephalai in 197 Philip advanced with the peltasts and the right of the phalanx leaving Nikanor nicknamed 'The Elephant' to follow with the rest of the army.[140] At the same battle the Roman army was divided into two wings.[141] Furthermore we have specific information that the Ptolemaic heavy infantry was divided into two halves, although this comes from prior to the infantry reform of the 160's. During the opening phases of the Fourth Syrian War (219-217 BC) the Ptolemaic phalanx was commanded and trained by Andromachos of Aspendus and Ptolemy son of Thraseas (Polyb. 5. 65. 4). Following the desertion of the latter to the Seleucid cause his place was taken by one Sosibios who was present at the battle of Raphia in 217 (5. 83. 3).

Only Asclepiodotus (3. 2) calls the whole infantry body a phalanx.[142] Aelian (9) and Arrian (11. 6-8) call it a *tetraphalangia*. Given the existence of this 'quadruple phalangia' and the 'double phalangia' (*diphalangia*) one suspects that the single *phalangarchia*, the equivalent of the Roman legion, may have alternatively been called a *phalangia* or phalanx. Although the standard Greek word used for the legion in literary texts is *tagma* a multitude of other Greek words are also used.[143] Perhaps the simple word phalanx itself was an acceptable alternative for *phalangarchia* in late Hellenistic usage. In the surviving fragment of his *Acies Contra Alanos* Arrian regularly uses the word phalanx qualified by a number to mean legion.

---

[139]Livy 37.40.1; Appian, *Syr.* 32; (Pydna) Livy 33. 4. 4.

[140]Polyb. 18. 24. 2; Livy 33. 8. 8 has 'wing'. For Nikanor see Paul Schoh, 'Nikanor (16)' *PW* xvii, 1 (1936) col. 270.

[141]Poly. 18. 23. 7; Livy 33. 8. 3.

[142]See Bruno Helly, *L'État thessalien. Aleuas le Roux les Tétrades et les Tagoi* (Lyon 1995) pp. 156-7 on the translation of this passage.

[143]Hugh J. Mason, *Greek terms for Roman Institutions (American Studies in Papyrology Vol. 13)* (Toronto 1974) p. 163.

To return to the Ptolemaic evidence once again, a papyrus dated to 29th July 127 mentions one Protarchos the *grammateus* (secretary) 'of the phalanx over which Polianthes holds command' (φάλαγγας ἐφ’ῶ̣ν̣ Πολιάνθης) is mentioned in line 7.[144] Noting that 'phalanx' is in the singular and that 'which' is in the plural, Wilcken proposed to emend phalanx to the genitive plural form (φαλάγγ<ων>).[145] Van't Dack, basing his suggestion on the *tetraphalangia* mentioned by Aelian and Arrian, suggested that the papyrus should perhaps be read with the number 4 in front of phalanx and he further proposed the insertion of an *iota* into phalanx, and to understand the text as stating that Polinathes commanded a unit called a '*tetraphalangia*' giving a text of (Δ) φαλαγγ<ί>ας. Though *tetraphalangia* is singular, it is quadruple in concept, which may have led the scribe to mistakenly put the relative pronoun in the plural.[146] Van't Dack's suggestion has much to recommend it, but in my view it requires too many changes to the text. Perhaps it would be best to leave the text as it is, with the word phalanx in the singular and to explain the grammatical inconsistency as being due to scribal error or confusion. If we leave phalanx in the text in the singular, it might be that in late Ptolemaic military usage the term phalanx could be used as term for a formation comprising a number of regiments, perhaps four, in the Ptolemaic army and perhaps numbering about 4,000 men. The Ptolemaic term 'phalanx' would thus be the equivalent of the Asclepiodotan *phalangarchia* and of the Roman legion.

We might now return to the honorific inscription to Kallikles son of Kallikles. Mitford restored Kallikles' final military title as καὶ εὐωνύμων [ἐπιστtά]τ̣η̣ν, understanding this title as being dependent on his post of *ilarchês* of the cavalry in Alexandria. Thus he conjectured that these would have been troops of 'the left wing' (εὐώνυμον κέρας) of the cavalry formation, which he thought might have been 'ancillary to the cavalry: some fancy formation, it may be, devised by this Instructor Royal in the Art of Tactics'. It might be that καὶ εὐωνύμων [ἐπιστtά]τ̣ην is rather a completely separate title which refers to his position as commander of the left wing of the Ptolemaic heavy infantry. Presumably each wing consisted of two phalanxes. Once again we should note that Asclepiodotus calls each wing of the heavy infantry either a 'double-phalanx' (*diphalangia*) or 'wing' commanded by a *kerarchês*.

Mitford's restoration of [ἐπιστtά]τ̣η̣ν is not entirely satisfactory, but it is difficult to suggest a suitable alternative. Logically the most suitable of the titles of rank listed by the writers of *Taktika* would be either[κεράρ]χ̣η̣ν or [διπαλαγγάρ]χ̣η̣ν but both of these are unsatisfactory for reasons of space. A more suitable restoration on grounds of space might be [κερατάρ]χ̣η̣ν, which term is given by Aelian (23) as the title of the commander of a 'wing' of elephants. All these restorations,

[144]Fr. Bilabel, *PBad* IV no. 47.

[145]Ulrich Wilcken, *Archiv für Papyrusforschung* 8 (1927) p. 88.

[146]Van't Dack, *Ptolemaica Selecta* p. 55.

however, run into the difficulty that we would expect the presence of the definite article τῶν before εὐωνύμω to give a meaning of 'and wing-commander of those on the right'. Thus a satisfactory restoration of this inscription defies us, given our current state of knowledge. I have not had an opportunity to examine the stone, and the suggestions offered above have only been made in the hope that some advance can be made in understanding the meaning of Kallikles' title.

Thus far we have been concerned only with evidence for organizational change in the Ptolemaic army, as so the evidence we have been dealing with has been from preserved literary, epigraphic and papyrological sources. We now turn to examine the archaeological evidence to see if there is any evidence for a change in the equipment used by the Ptolemaic infantry which might be associated with the infantry reform carried out in the 160's. The overwhelming bulk of representational evidence for the Ptolemaic army consists of the painted tombstones from the cemeteries of Alexandria which are almost entirely dateable to the late fourth and third centuries. Thus we have less evidence to deal with than one might expect.

Section 1 i

The Stele of Dionysios the Bithynian

Bibliography: E. Breccia in *BSAAlex* 15 (1914/5) p. 59-60 no. 5; *SB* 6241; M. Rostovtzeff, *The Social and Economic History of the Hellenistic World*[2] (Oxford 1953) I p. 288, pl. 37, 1; *PP* 2882; Blanche R. Brown, *Ptolemaic Paintings and Mosaics and the Alexandrian Style* (Cambridge Mass. 1957) p. 28 no. 27, pl. xx, 1.

Alexandria Museum no. 20919;

The painted tombstone of Dionysios the Bithynian (Figure 9), in the form of a naiskos with pediment, supported by two pillars, was found at an unknown date in the necropolis of Gabbari in Alexandria. It is 0.55 m high and 0.32 m wide. On the architrave is an inscription Διονύσιος Βιθυνός Χαῖρε. Brown describes this stele as 'full fourth style' which would place it within the second century.[147] She com-

[147]p. 32 'With Dionysius the Bithynian the full Fourth Style is reached. Lifelikeness also is gone. The abstract outline has reinstated itself. The drawing is shorthand and stereotyped. There are areas of colour but scarcely a pretence of modelling. The two figures stand repetitively one next to the other, with no noticeable interrelationship. The effect is of formula-making of the most naïve and simple sort.'

Figure 9. Stele of Dionysios the Bithynian (after Brown, *loc. cit.*).

pared the fourth style to the Sidon stelai,[148] therefore it might be possible to regard this as one of the latest painted stelai from Alexandria showing soldiers, and it might therefore show an infantryman dating to after the reform. Breccia first briefly described the discovery of this stele, mistakenly speaking of it as showing a horseman. Consequently Dionysios has gone into the academic literature as a cavalryman.[149]

Two figures are shown, the larger being the deceased Dionysios himself while the smaller represents an attendant. The background is blue and the outlines and skin of both figures are brown. Dionysios is depicted wearing dark brown ankle-high boots, a wreath around his head, and a dark brown tunic with a vertical stripe down the middle. His attendant wears a white tunic, a blue item of headgear, presumably the iron helmet belonging to his master, and carries a 'yellowish' shield. The shield is presumably white but with yellowish tints used in the painted modelling of the contours of the shield.

Rostovtzeff describes the spear carried by Dionysios as being probably a *sarisa*, but the head is shown at the top of the field and so it seems that Dionysius carries a six-foot fighting-spear, while his attendant carries a pair of javelins. The spear carried by Dionysios could be regarded as being the equivalent of the Roman *hasta* and the two javelins carried by his attendant as being the equivalent of two *pila*. As Polybius (6. 23. 16) describes the Roman army only the *triarii* carried the *hasta* while the *hastati* and *principes* carried a pair of *pila*, but Ptolemaic practice does not need to have been identical. Certainly the practice of carrying both a fighting spear and a pair of throwing spears is foreign to Hellenistic military tradition, and I know of no other representation of a Hellenistic warrior showing this combination of weapons. The *thureos* shield is similarly the equivalent of a Roman *scutum*. The fact that Dionysios wears boots is also unusual. Greek hoplites are generally shown barefoot and the boots worn by Dionysios might be regarded as the equivalent of Roman *caligae*. The helmet worn by the attendant is too indistinctly shown to justify commentary. The stele does not show Dionysios as carrying a sword, but the stele is heavily damaged in the area where his waist is shown and it may have originally been shown there. It is also possible that the attendant might be carrying his sword.

The absence of a cuirass of any sort deserves some comment. According to Polybius (6. 23. 14-15) when the Roman army was mobilized only those soldiers with property of more than 10,000 drachmas were required to equip themselves with a mail cuirass, the rest wore a small square bronze plate over the heart. Thus

---

[148] p. 46 Dionysios belongs to the stratum who by the second century were buried in loculi. The Fourth Style represents 'a drop into the stratum of popular art' and loses connection with high art. 'It is exactly such a drop into the popular stratum, with its characteristic freezing of motifs and crudeness of rendering, which also appears in the painted loculus slabs of Sidon'.

[149] Launey I p. 435; II p. 1211.

the mail cuirass only became general throughout all the maniples of the legion after 123 BC, when a law was passed during the first tribunate of C. Gracchus providing for the free distribution of arms to every citizen-soldier in the army at public expense.[150] Another reason why Dionysios is not shown wearing a cuirass may be that the cuirass was not normally worn by troops on garrison-duty, and consequently has not been shown by the artist. As a general observation one might add that Hellenistic painted funerary stelai showing soldiers tend to show them wearing 'walking out dress' rather than 'battle dress'. [151]

Section 1 j

The Tombstone of Salmas of Adada

Bibliography: Gustave Mendel, *Musées Impériaux Ottomans. Catalogue des sculptures grecques, romaines et byzantines, 1* (Istanbul 1912) pp. 264-6 no. 105 (with earlier bibliography); A.H. Sayce, *Classical Review* 28 (1914) 197 no. 2; Klaus Parlasca, *Syrische Grabreliefs hellenistischer und römischer Zeit. Fundgruppen und Probleme* (= *Trier Winckelmannsprogramme 3,* 1981) pp. 6, 23 n. 15, pl. 2, 2; Petros Dintsis, *Hellenistische Helme* (Rome 1986) I pp. 144, 147, 290 no. 248, II pl. 68, 1; Bezalel Bar-Kochva, *Judas Maccabeus. The Jewish Struggle against the Seleucids* (Cambridge 1989) 581 pl. ix.

Istanbul, Archaeological Museum (Inv. 1167).

The painted loculus-slabs from Sidon constitute an important group of archaeological evidence bearing on the equipment and uniform of Hellenistic armies. Their date, and in consequence their assignment to either the Ptolemaic or the Seleucid army is an extremely contentious point. In Appendix C I have attempted to demonstrate the probability that they were deposited by a Ptolemaic garrison during the years 147-145 BC. The loculus slabs were in the form of stelai, used to seal the entrance to the loculus. Of the 17 recovered from Sidon 5 were too damaged to allow any certainty as to what type of troop was shown in the painting, while 4 show archers, cavalrymen or peltasts. The remaining 8 show infantrymen equipped

---

[150] Plut., *Vit. C. Gracch.* 5. 1; On this passage see Emilio Gabba, *Republican Roma, The Army and the Allies* (Oxford 1976) 6-7.

[151] Cf. for example the painted tombstones from Pagasai-Demetrias in Α.Σ. Αρβανιτοπουλου, *Γραπται Στηλαι Δημητριαδος-Παγασων* (Athens 1928) *passim.*

Figure 10. Tombstone of Salmas from Sidon.

with *thureoi*. One of these 8, the loculus slab of Salmas son of Moles of Adada, is of particular interest as it shows a soldier who also wears a mail cuirass. Thus Salmas at least is seemingly serving in an infantry unit of the Ptolemaic army which had been equipped in the Roman manner. Presumably the soldiers shown on the other 7 loculus slabs, who are equipped in an identical manner to Salmas but without the cuirass, also belong to units of the Ptolemaic army equipped in the Roman manner. The same comments made with respect to the stele of Dionysios, that not all Roman infantry equipped themselves with the mail cuirass during the second century, and that in general troops preferred not to be shown wearing their full panoply on funeral paintings, also apply here. Nevertheless we shall restrict our discussion to the loculus slab of Salmas (Figure 10).

The inscription has proved difficult to restore, but it now stands as follows.

Σάλμα Μολ[εω 'Α]δαδε[ῦ]
χρησ[τὲ χαῖρε].

Salmas son of Moles of Adada,
good man farewell!

The damaged inscription of the stele of Salmas was first restored by Sayce, with a Semitic name and patronymic, as Salmamodes son of Hadades. Louis Robert recognized that the deceased came from the Pisidian city of Adada, and restored the Anatolian personal name Salmas.[152] He subsequently demonstrated that the patronymic, which must be restored with two letters only following the break, can be restored with an irregular genitive form of the Anatolian personal name Μολης.[153] Clearly the restoration of Sayce was influenced by the fact that he thought a Semitic personal name would be appropriate for a Seleucid soldier, while the restoration of Robert is in line with the marked presence of Anatolians among the deceased.

The helmet worn by Salmas is of the same type as that worn by all the other infantrymen equipped with *thureoi* on the Sidon stelai, though this is not immediately obvious as the artist has found difficulty in depicting the peak frontally. The crest-box is not clear due to minor damage to the stele, but it can be definitely stated that the helmet is not of the Boeotian type, as has been suggested by Bar-Kochva. Dintsis, who dated the stele to the first half of the second century BC, classified Salmas' helmet as a form of 'Glockenhelm', or 'Bell-shaped helmet'. Note that the front of the peak of the helmet comes to a point, a feature which is only shown

---

[152] L. Robert, *Opera Minora Selecta* I (Amsterdam 1969) p. 187; L. Zgusta, *Kleinasiatische Personennamen* (Prague 1964) 1360-1.

[153] Zgusta, *op. cit.* 946-1; L. Robert, *Noms indigènes dans l'Asie-Mineure gréco-romaine* (Paris 1963) 353-5.

clearly in this stele.

The mail cuirass is shown without shoulder guards. It appears to be almost like a T-shirt in shape, though without any sleeves. Rectangular in shape, without any reinforcement at the shoulders, it has a simple neck-hole at the top to put the head through. The stucco has been badly damaged around the waist, so we do not know whether Salmas is wearing a waist-belt or not. It is highly probable that he is though, not only because of the way the cuirass is shown narrowing towards the waist, but also because it is always standard practice to support the mail cuirass with a waist-belt in order to take some of the strain imposed by the weight of the cuirass away from the shoulders. The outline of the baldric cannot be clearly made out on the stele. A single brown line can be made out, as if Salmas is wearing a simple leather baldric like the other infantrymen on the stelai. There seems to be another dark line running parallel to this brown one, indicating that the belt may be much wider and covered in iron plates in the Roman manner. The boots are also different from those worn by the other infantrymen on the stelai. They seem to be a proper 'boot' rather than a strap-work sandal. There is no visible lacing, and it looks as though the boots are laced at the front, but the laces are concealed in a 'pocket' running down the front of the boot. The top of the boot is painted in a rather lighter shade of brown, as if the felt 'sock' normally worn beneath the sandal is protruding above the boot proper. Note that there is a slight indentation in the toe of the boot between the big toe and the rest of the foot.

Section 1 k

The Itanos Stele

Figure 11 shows a funerary stele found in the sea near Itanos and now in Aghios Nikolaos Museum.[154] The coast of eastern Crete has sunk, and it would be a reasonable assumption that this stele, brought up from the sea-bed in fishermen's nets, would originally have come from a cemetery used by the Ptolemaic garrison stationed at Itanos. Whether the individual shown is a Ptolemaic soldier or not, he seems to be equipped in the Roman manner. Although the stele is heavily eroded, one can make out that the soldier wears a crested helmet. In his right hand he holds a medium-length fighting spear and his left hand rests on a *thureos* which rests on the ground. On his feet he wears boots, a typical feature of Roman infantry equipment. Over a sleeveless tunic he wears a cuirass. The details of the cuirass are too eroded to make out much detail, so it is not possible to decide whether it is a mail

[154] *Kret. Khron.* 23 (1971) p. 498; *Arch. Delt.* 26 (1971) B 2 p. 500-1 pl. 517d; *BCH* 100 (1976) p. 723, p. 727 fig. 339.

Figure 11. Funerary stele from Itanos (photograph: Aghios Nikolaos Museum).

cuirass or a muscle cuirass. It probably has two shoulder-guards and perhaps some sort of belt at the waist. Under the left shoulder one can make out the hilt of a sword. This is not a Roman feature, but it does not affect the image of this warrior as one equipped in the Roman manner. The style of the stele would fit a date in the second century. Lembesis dated the stele to the last decade of the second century.

Itanos was a Ptolemaic protectorate for much of the Hellenistic period, although the details of this protectorate are unclear. We have already mentioned an inscription from Itanos dating to the reign of Ptolemy IV Philopator (217-209 BC) left by the Roman *phrourarchos* Lucius son of Gaius, which demonstrates that there was a Ptolemaic garrison stationed in the city.[155] A second *phrourarchos* Philotas son of Genthios, an Epidamnian, is mentioned in an inscription dateable to the beginning of the second century, which Guarducci would date to the reign of Ptolemy V Epiphanes (205-181 BC) though others have sought to date it to the reign of Philometor.[156] As the inscription names Philotas as a *chiliarchos* as well as *phrourarchos*, it might be preferable to place this inscription before the infantry reforms of the 160s. It is possible that the garrison was withdrawn at some point late in the reign of Epiphanes or early in the reign of Philometor, but it was probably subsequently re-introduced during the reign of Philometor at some point after the year 164/3.[157] An Itanian inscription mentions the *prostasia* of King Ptolemy (Philometor).[158] This word is generally interpreted as referring to the Ptolemaic protectorate over Itanos.[159] The word *prostates* is, however, also used of the position of presidency of the Cretan *Koinon* to which Philip V was elected in 217,[160] so it seems preferable to interpret the word *prostasia* as referring to Philometor's presidency of the Cretan *Koinon*. After the end of the Third Macedonian War Knossos appealed to Philometor, then in alliance with Gortyn, to mediate in the internal war in Crete.[161] It was probably as a result of his successful mediation, which brought peace to the island, that Philometor was elected *prostates* of the Cretan *Koinon*. The garrison was probably permanently withdrawn in 145 following the death of Philometor. For this reason a date between the 160's and 145 would seem most suitable for this stele.

---

[155] *ICret.* III, iv, 18 (p. 115-6).

[156] *ICret.* III, iv, 14 (pp. 112-4).

[157] See Guarducci in *ICret.* III p. 77; Stylianos Spyridakis, *Ptolemaic Itanos and Hellenistic Crete* (= *University of California Publications in History, Volume 82,* Berkeley Los Angeles London 1970) p. 82-3.

[158] *ICret.* III iv 9, 107 (p. 100).

[159] Eg. Walbank, *Commentary* II p. 59.

[160] Polyb. 7. 11. 9; Guarducci, *Riv. Fil.* n.s. 16 (1938) pp. 53-4; Sheila L. Ager, *JHS* 114 (1994) p. 1 n. 2.

[161] Henri Van Effenterre, *La Crète et la monde grec de Platon a Polybe* (*Bibliothéque des Écoles françaises d'Athénes et de Rome Fasc. 163,* Paris 1948) pp. 216, 266-7.

Section 11

The Malcove Fragments

Bibliography: G.M.A. Hanfman, 'New Fragments of Alexandrian Wall painting' in *Alessandria e il mondo ellenistico-romano, studi in onore di Achille Adriani 2*, = *Studi e materiali, Istituto di archeologia Universita Palermo* 5 (Rome 1984) pp. 242-255, tav. xliv-xlvi; Marylin McKay in ed. Sheila D. Campbell, *The Malcove Collection. A Catalogue of the Objects in the Lillian Malcove Collection of the University of Toronto* (Toronto Buffalo London 1985) pp. 33-5.

Toronto University M82.291.

Some twenty fresco fragments mounted in four modern panels were purchased for the Malcove Collection and subsequently bequeathed to Toronto University. They are said to come from Alexandria and date to the Late Antique period. Odysseus is shown in panel 1, so perhaps all, or at least some, of the fragments belong to frescoes showing scenes from the Odyssey.

Panel 2 (Figure 12) consists of three fragments. The central fragment shows a golden crater on top of a simplified Corinthian column. The left-hand fragment shows an armed figure in a yellow helmet surmounted by a crimson plume fixed in a white plume-holder. He wears a white cuirass decorated by painted black lines with a violet shoulder-guard. A red cloak is shown at the shoulder. The right-hand fragment also shows an armed figure dressed in a blue tunic with a red border shown round the neck. The cuirass is yellow with red shown on the belt. The helmet and sword-hilt are white. The shield is shown in yellow. Not enough of the shield is preserved to enable us to establish its shape. Presumably it is of the *thureos* type like all the others shown in the panels, but it could possibly be of another shape.

Panel 3 (Figure 13) consists of four fragments all from the same fresco. A total of seven guards are shown standing in front of a building. All guards carry medium-length fighting spears and *thureos* shields. In some cases the nails attaching the metal rim to the body of the shield are clearly shown. They all wear helmets with extended cheek-pieces. In some cases the ridge-crest of the helmet is clearly shown, and all helmets are plumed. The outer left fragment shows guards numbers one and two standing in front of the black corner of a building shown in the back-

Figure 12. Malcove fragments, Panel 2 (photograph: The University of Toronto Malcove Collection).

ground between their heads. The inner left fragment shows a further two guards standing in front of a tall white masonry structure with eight visible courses flanked by trees. Guard number three carries a white shield with a pink rim and his helmet has a black plume. Guard number four has a red shield with a white rim with a black central spine. His arm is clear in this single example, which makes it plain that the guardsmen all wear tunics leaving the arm bare below the shoulder. Behind him two rectangular structures are shown, seen in elevation but laid out flat as if viewed from above. Campbell noted that 'the architectural drawing convention certainly suggests that it was made in Egypt'. The left inner fragment shows guards number five and six standing in front of five containers filled with a red substance, perhaps wine. The left outer fragment shows the head and shoulders of a single guard number seven standing in front of a vertical brown pier to his left. The helmet of guard number seven is shown most clearly. It closely resembles the helmets shown on the majority of the Sidon painted loculus slabs, for example on that of Salmas.

Panel 4 (Figure 14) shows ships filled with troops and troops fighting on the shore. The large fragment upper left shows part of a ship sailing in a sea painted blue-black. On board two soldiers are shown armed with medium-length fighting spears, helmets, and *thureos* shields and wearing belted tunics decorated with stripes. In the next fragment to the right is a feature indicated by a criss-cross pattern, which Hanfmann identified as a canvas deck cabin, with two indistinctly painted soldiers standing in front of it. The large fragment upper right shows parts of two

Figure 13. Malcove fragments, Panel 3 (photograph: The University of Toronto Malcove Collection).

ships. Two soldiers with spears, helmets and *thureoi* and two round bronze shields attached to the gunwale are shown on the upper ship and more soldiers armed in the same way on the lower boat. In the lower fragments of the Panel armed figures fight each other. In all cases the helmets and *thureoi* are of the same type. In the large central lower fragment a warrior is shown wearing a red tunic decorated with white stripes running vertically down from each shoulder and along the arm-hole of the tunic. His helmet, sword and baldric are also shown in white. In his right hand he holds a medium-length fighting spear and in his left a *thureos*. The fragment to the right shows a figure wearing a helmet with curving cheek-pieces shown in white.

Hanfmann suggests a date of late third or early fourth century AD while McKay dates panel 1 to circa 400 and the others to between 400 and 500 AD. Though the basic outline remains the same, some organizational changes have taken place. Hanfmann (p. 248) has pointed out that the helmets and cuirasses are of a late Classical-Hellenistic type, and though the frescoes are Late Antique, they are clearly influenced by Hellenistic models. The frescoes are in the Late Classical - Early Hellenistic panel picture tradition of large figures (megalographic) and more especially in the Late Hellenistic tradition of panoramic illustration from a high viewpoint, as demonstrated by the Esquiline series of Odyssey paintings. 'The first adaptations of these traditions took place in Hellenistic Egypt as proved by the survival of Egyptian architectural schemes and by the Hellenistic armour of the warriors. In my opinion this happened in the middle or late Hellenistic period in the royal studios at Alexandria' (p. 255).

Hanfmann (p. 249-50) argues that the building shown in Panel 3 could be a fortress, or a fortified palace, or a sacred structure of Egyptian type, but because of the row of armed men standing in front of it, a fortress-palace seems more likely. As no situation in the Odyssey quite fits, and Hanfmann rejects any identification with the palace of Alcinous, he suggests that the Iliad or any of the other post-

Figure 14. Malcove fragments, Panel 4 (photograph: The University of Toronto Malcove Collection).

Homeric epics could have formed the subject of such a scene. On the other hand Hanfmann insists that a historical scene of the Ptolemaic era cannot be completely ruled out, noting that the guards wear Hellenistic armour. This, along with the use of Egyptian building conventions, argues a Hellenistic model, although he thought the poses were influenced by Late Roman models.

It could be argued that the military figures shown are based on Ptolemaic models derived from the equipment of the Ptolemaic army after the infantry reform of the 160s. It hardly needs to be stated that infantry equipped with medium-length fighting spears and *thureoi*, such as are shown on the Sidon loculus-slabs, are only typical after the reform. The helmet of guardsman number seven is particularly close to those worn by the majority of the soldiers shown on the Sidon tombstones.

The best indicator of the date of the original models of the warriors shown on the Malcove fragments is the sword-hilt shown in the right-hand fragment of Panel 2. This type of sword, ending in a multi-lobate (in this case bilobate) pommel, is of a distinctive Hellenistic type. It is derived from Celtic prototypes first imported into the eastern Mediterranean by the Galatians. Such swords are shown in the weapon friezes decorating the balustrade of the Temple of Athena Nikephoros in Pergamon. The date and of the latter monument is not known for certain, but the most plausible suggestion is that it was erected to commemorate Pergamene victories over Seleucid and Galatian forces in the years 230-229.[162] The swords of this type

[162]Cf. A.W. Lawrence, *Later Greek Sculpture and its Influence on East and West* (London 1927) pp. 3, 20.

Figure 15. Stele of Dioskourides of Balboura (Archaeological Museum, Istanbul).

shown on the balustrade sculptures may well be booty taken from the Galatians. Where shown, the swords used by the soldiers of the Sidon tombstones are of this type, as is most clearly shown in the tombstone of Dioskourides of Balboura (Figure 15).[163] It could be argued that swords with multi-lobate pommels had only spread from Galatia to other areas of Anatolia, and it is for this reason that they are

[163]Gustave Mendel , *Musées Impériaux Ottomans. Catalogue des sculptures grecques, romaines et byzantines 1* (Istanbul 1912) no. 102 (with previous literature); A. H. Sayce, *Classical Review* 28 (1914) p. 197 n. 3; Paul Couissin, *Les Institutions militaires et navales* (Paris 1932) pl. 38, 2 (drawing); Blanche R. Brown, *Ptolemaic Paintings and Mosaics and the Alexandrian Style* (Cambridge

carried by the Anatolian *symmachoi* of Ptolemy Philometor at Sidon. Swords of this type are, however, also attested in late Hellenistic Egypt in the form of miniature clay models of swords with multi-lobate pommels (Figure 16). Many examples of these miniature terracotta swords come from Naucratis.[164] It is therefore certain that swords of this type were used in late Ptolemaic Egypt.

It could be argued therefore, that the Malcove panels are derived from late Ptolemaic models showing soldiers equipped in the same manner as those on the majority of the Sidon tombstones. Needless to say, this argument is rather circular, and the paintings cannot therefore be taken as evidence in support of the infantry reforms. Nevertheless they are of sufficient interest to be included in this discussion.

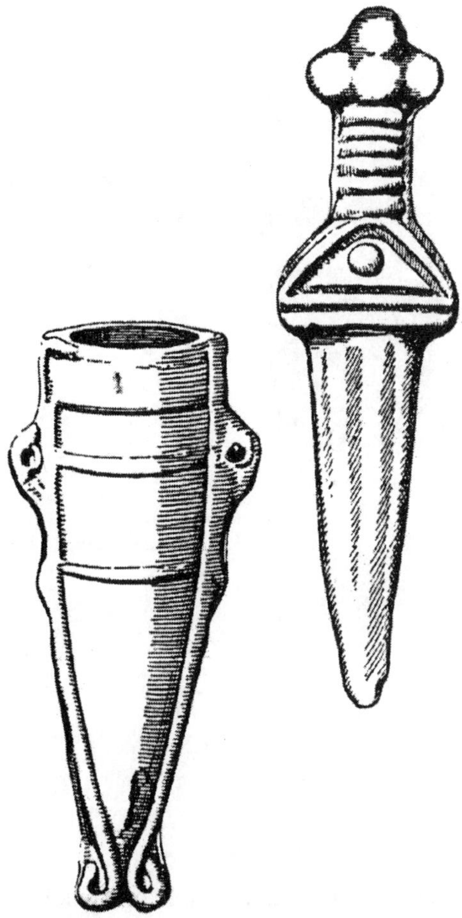

Figure 16. Clay model of sword, probably from Naucratis (after Walters, *op. cit.* p. 259 fig. 53).

Mass. 1957) p. 87, pl. 20, 2; Volkmar Von Graeve, *Der Alexander Sarkophag und seine Werkstatt* (=*Istanbuler Forschungen 28,* Berlin 1970) p. 89 n. 47; Petros Dintsis, *Hellenistische Helme* (Rome 1986) pl. 41, 3; Klaus Parlasca, *Syrische Grabreliefs hellenistischer und römischer Zeit. Fundgruppen und Probleme* (= *Trier Winckelmannsprogramme 3,* 1981) pp. 6, 23 n. 14, pl. 2,3; Bezalel Bar-Kochva, *Judas Maccabeus. The Jewish Struggle against the Seleucids* (Cambridge 1989) p. 582 pl. x.

[164] H.B. Walters, *Catalogue of the Terracottas in the Department of Greek and Roman Antiquities, British Museum* (London 1903) pp. 259-60 C 628-632. For a later, Roman, example of one of these Egyptian miniatures with a later shape see Valdemar Schmidt, *De Graesk-Aegyptiske Terracotter I Ny Carlsberg Glyptothek* (Copenhagen 1911) pl. LV, fig. 165 (E 787) = Mette Fjeldhagen, *Graeco-Roman Terracottas from Egypt* (Ny Carlsberg Glyptotek, Copenhagen 1995) p. 206 no. 210 (with references to further examples).

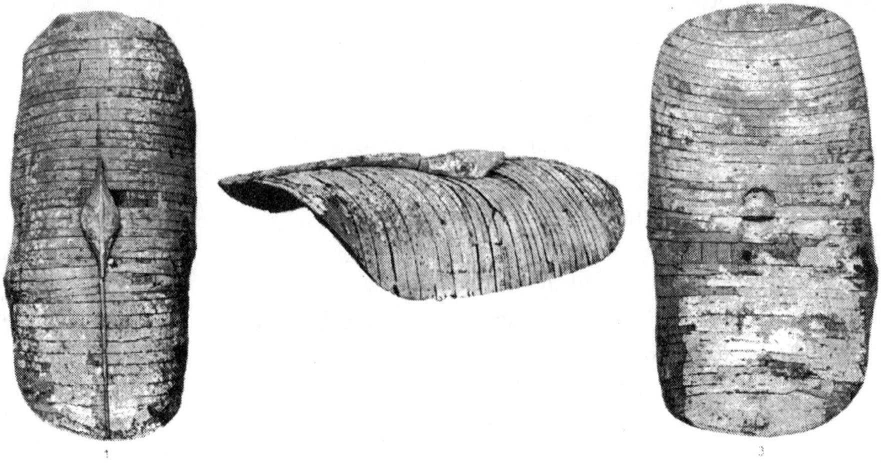

Figure 17. The Kasr el Harit Shield (after Kimmig, *op. cit.* pl. 23).

Section 1 m

The Kasr el Harit Shield

Finally we turn an example of a shield of Roman type found in Egypt which was fully published by Kimmig (Figure 17).[165] It was found in 1900 at Kasr el Harit in the Fayoum during searches for papyri and is now in the National Egyptian Museum in Cairo. It was found in a section of an ancient cemetery in which buildings were mixed among the graves. The shield was found in one of these buildings. The buildings all seemed to date to the late Ptolemaic period, and one contained papyrus fragments dateable to the late second century BC.[166] The shield would best fit a date in the late second century, as it is closest in size and shape to the shields used by the infantry shown on the 'Domitius Ahenobarbus Monument' from Rome now in the Louvre (Figure 18), which has been dated to the consulship of Cn. Domitius Ahenobarbus in 115 BC.[167]

Kimmig gives an admirably detailed description of the shield, supported by a diagram which is reproduced here as Figure 19. It is 1.28 m tall and 0.635 m wide.

[165]Wolfgang Kimmig, 'Ein Keltenschild aus Äegypten', *Germania* 24 (1940) pp. 106-111, pl 23.
[166]B.P. Grenfell, A.S. Hunt, D.G. Hogarth, *Fayûm towns & their Papyri* (London 1900) p. 58.
[167]Mario Toarelli, *Typology and Structure of Roman Historical Reliefs* (1982) p. 15.

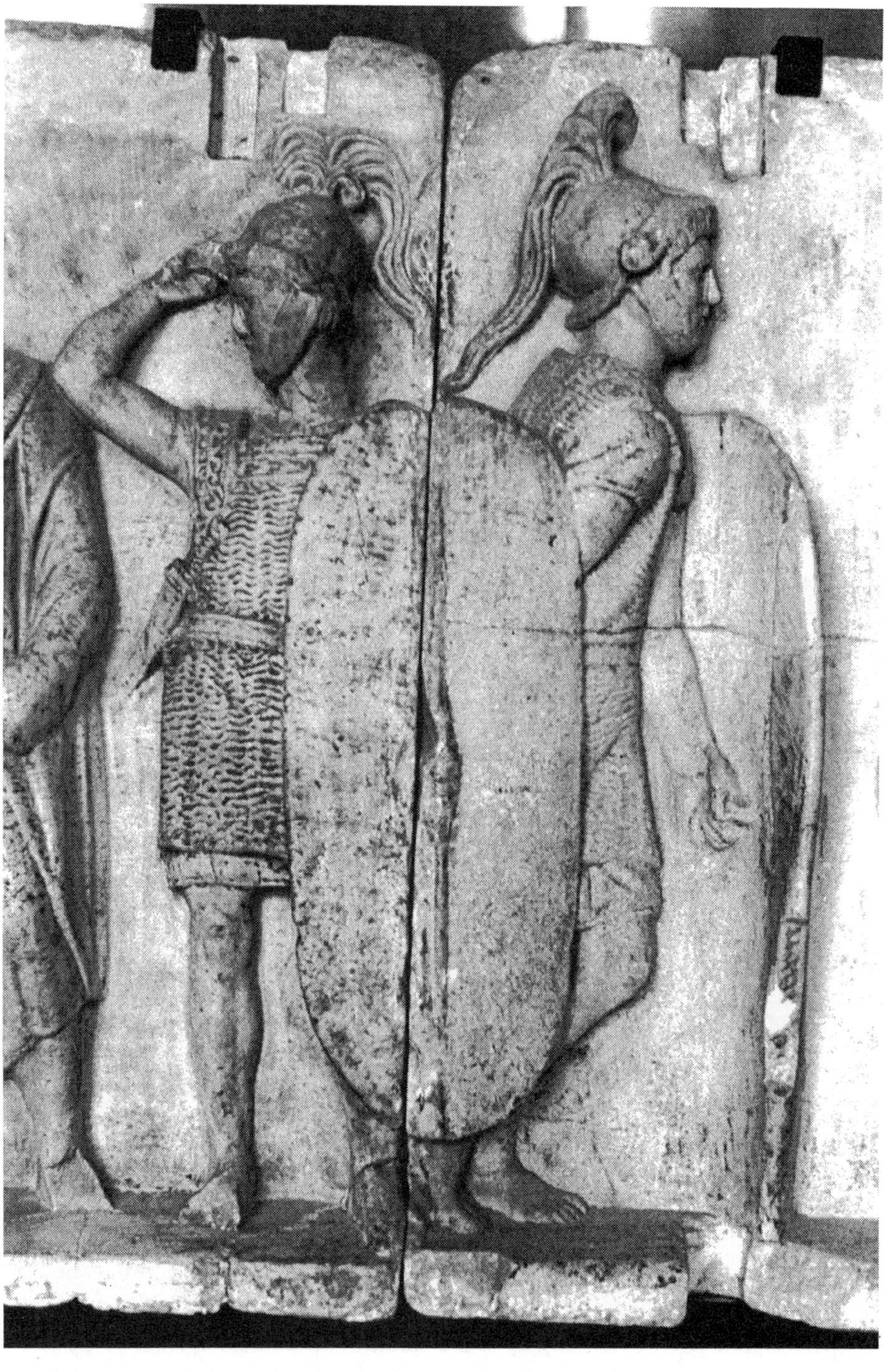

Figure 18. Infantry shown on the Domitius Ahenobarbus Monument in the Louvre, Paris (author).

Figure 19. Section of the Kasr el Harit shield (after Kimmig, *op. cit.* p. 108 fig. 1).

The shield is made of wooden laths in a triple-ply construction. The central layer consists of nine or ten laths running lengthways, each with an average width of 6-10 cm. On the inside and outside of this central layer, and presumably glued to it, were a layer of 40 narrower laths, with a width varying between 2.5 and 5 cm, running cross-wise at right-angles to the laths of the central layer. The laths were all about 2-3 mm thick, and the shield as a whole had a thickness of no more than a centimetre, although it was stronger and thicker in the middle portion where the lengthways laths were somewhat thicker. The edges of these vertical laths were slightly carved at an angle to give the shield its partly cylindrical curvature. The handle, *umbo* and *spina*, all also wooden, were attached to the main body of the shield with nails. Both sides of the shield were finally covered with a layer of woollen felt. The inside layer overlapped the outside layer by 50-60 mm, giving it a 'rim'.

In size and method of construction the Kasr el Harit shield corresponds closely with the Roman shield as described by Polybius (6. 23. 2-5). As has already been mentioned, it is also close in shape and size to representations of Roman shields, such as the Domitius Ahenobarbus Monument. There can be little doubt that we are dealing with a shield of Roman type. Given the date of the shield, it cannot have been deposited by a Roman soldier, as no Roman units entered Egypt before the middle of the first century. Furthermore *thureoi* used earlier in the Hellenistic world, based on Galatian shields, were generally smaller and narrower than the Roman shield, more oval in shape, and without the partly cylindrical curvature. The easiest way to explain the presence of this shield in the Fayoum is that this type of shield was in general use in the Ptolemaic army in the late second century BC. The complete absence of any metallic elements in the shield - rim, *umbo* or *spina*, is unexpected, but we presumably have the shield in its final state, abandoned by its cleruch owner, rather than an unfinished example. Whether the shield was imported or manufactured in Egypt is a matter of speculation. It is perfectly possible that the latter was the case, since no suitable wood for shields grows in Egypt, the wood would have to be imported anyway. Pliny (*HN* 16. 209) tells us that the most suitable woods for making shields are those in which an incision causes the wood to draw together at once and close its own wound, these include vine, willow, lime, birch, elder and both kinds of poplar. Kimmig thought the Kasr el Harit shield was made of birch.

Section 1 n

Conclusion

The archaeological evidence, including the representational evidence and the Kasr el Harit shield, show that the Ptolemaic infantry was re-equipped in the Roman manner at some point during the Ptolemaic period. The epigraphic and papyrological evidence shows that the maniple was introduced into the Ptolemaic infantry at some point before the end of the triple reign of King Ptolemy, King Ptolemy the Brother and Queen Cleopatra, and certainly before 163/2. Presumably Roman tactics and equipment were adopted on this same occasion as well as Roman organization.

CHAPTER 2

EVIDENCE FOR REFORM IN THE SELEUCID ARMY

We shall now examine the evidence for military reform in the Seleucid Army. We have much less documentary or archaeological evidence available for research into the Seleucid state than is the case with Egypt, so it comes as no surprise that we shall be dealing with only two passages of text and one piece of archaeological evidence. We begin with the literary evidence.

Section 2 a

The Daphne Parade

In 166 Antiochus IV held a magnificent parade at the town of Daphne followed by 30 days of festivities in the city of Antioch. Daphne lay eight kilometres from Antioch, allegedly on the site where the nymph Daphne, pursued by Apollo, was turned into a bay-tree. Daphne housed the important sanctuary of Pythian Apollo, the protecting deity of the Seleucid dynasty, and was selected as the site for the parade because of the ritual games of Apollo which were held there.

The only description of the parade which has survived comes in a single manuscript. This manuscript preserves the text of the *Deipnosophistai* of Athenaeus of Naucratis, which contains a fragment of Polybius (conventionally numbered 30. 25) describing the Daphne Parade. The fullest piece of text in Athenaeus (5. 194 c - 195 f) preserves an account of the parade and festivities, which is sufficiently detailed to leave the impression that the paraphrase of Polybius made by Athenaeus is reasonably complete. A second fragment in Athenaeus (10. 439 b- d) repeats some of the later narrative of the principal passage dealing with the festivities, though in slightly less detail. Diodorus (31. 16. 2-3) similarly preserves a partial description of Antiochus' role in supervising the parade and in the festivities after, which is useful in being separately derived from Polybius, and thereby demonstrating how faithful our excerpts are to the original.

The ultimate source of Polybius' information is unknown. We only have detailed descriptions of two Hellenistic royal parades preserved in the literary sources, the first being the Daphne Parade itself, and the other a description of the parade of Ptolemy II Philadelphus held in Alexandria in 275/4 BC, contained in a fragment of Kallixeinos of Rhodes's book on the city of Alexandria preserved in Athenaeus 5. 196 a - 203 b.[1] In her book dealing with this parade Ellen Rice has noted that even if Kallixeinos was himself an eyewitness or used an eyewitness source, he would still have been dependent on some source of information on the numbers, weight, size and cost of the objects that appeared in the parade. She notes that Kallixeinos mentions αἱ τῶν πεντετηρίδων γραφαί 'the *graphai* of the penteric festivals' (Athen. 197 d). Just what these *graphai* may have been is open to dispute. Rice notes that the formal nature of the *graphai* is implied in the phrase: 'The normal interpretation of γραφή is a visual representation of some kind, usually of a drawing, painting or picture, but it can have a connotation other than visual'. Rice considers it possible that Kallixeinos could have used a written source. Jacoby (*FGrH* 627 ad loc.) even suggested an emendation of γραφαί to ἀναγραφαί to convey the meaning of a written document, but Rice notes that 'this is unnecessary if the words can be synonymous'. She then considers the possibility that Kallixeinos may be referring to the Royal Archives, or *basilikai anagraphai* and notes that 'it is difficult to explain how he could have known so many of the visual details he records so vividly if not from some visual source'. In conclusion she thinks we may be dealing with manuscript illustrations, but she considers a range of other possibilities.[2]

The same comments as to the sources at the disposal of Kallixeinos are also true of Polybius' description of the Daphne Parade. It is surely probable that Polybius made use of a contemporary written description, perhaps with accompanying illustrations, put into circulation by the monarch responsible, in this case Antiochus IV, in order to enhance the propaganda value of the parade. We know that this happened with court processions in the early modern period.[3] At this time Polybius

---

[1] Victoria Foertmeyer, 'The Dating of the Pompe of Ptolemy II Philadelphus' *Historia* 37 (1988) pp. 90-104 esp. 103 (date).

[2] E. E. Rice, *The Grand Procession of Ptolemy Philadelphus* (Oxford 1983) pp. 171-4.

[3] Eg. Roy Strong, *Art and Power, Renaissance Festivals 1450-1650* (Woodbridge Suffolk 1984) p. 47 on the practice of issuing an 'Illustrated Royal Entry' writes 'one of the objects of such publications was to reinforce by means of word and image the central ideas that motivated those who conceived the programme'. On Renaissance royal processions see also Rainer Roy and Friedrid

had begun his internment in Rome where he built up a wide circle of friends, including the exiled Seleucid prince Demetrios who would have been kept informed of contemporary matters in Syria.

Athenaeus preserves two fragments of a work by one Protagorides of Cyzicus entitled *Concerning the Festivals at Daphne* (Περὶ τῶν ἐπὶ Δάφνῃ Πανηγύρεῶν).[4] Abel suggested that Protagorides was an official panegyrist appointed by Antiochus.[5] While this is perfectly possible it is impossible to prove. The two fragments preserved are remarkably uninformative. The first fragment tells us that a third kind of dinner is the Egyptian, where no tables are placed beside the guests, but boards are carried around among them. The second tells us that in the second book Protagorides says that a variety of musical instruments were played. Consequently the surviving fragments give us no clues as to what type of work it was or when it was written. The title suggests that the work was of a general nature, perhaps dealing with the historical development of the festivals at Daphne. We should note that the title of Protagorides' work is in the plural, whereas the parade at Daphne in 166 was a single festivity, forming only the opening event of a series of spectacles that lasted 30 days (Polyb. 30. 26. 1). The succeeding events took place at Antioch rather than Daphne. On balance it is unlikely that Polybius used Protagorides as a source, although the possibility cannot be excluded completely.[6]

The purpose and date of the Daphne Parade are the subject of some dispute. Although these questions are not of central relevance to the subject of this book, they are of sufficient importance to warrant discussion. They have, however, been relegated to Appendix D rather than being dealt with in the main body of the text.

Section 2 b

The Roman-Armed Infantry in the Daphne Parade

The Daphne Parade started with a vast military parade in which all the forces of the Empire were assembled. Many of the festivals held by the Greek *poleis* or the Hellenistic kingdoms had a military aspect, but the Daphne Parade seems to be

---

Kohler, 'Festaufzug, Festeinzug' in *Reallexicon zur deutschen Kunst-Geschichte* Vol. 8 (1987) 1417-1520. I thank Richard Brzezinski for acquainting me with this last work.

[4] Athen. 4. 150 c, 4. 176 a; *FHG* 4. 484; .

[5] Abel, *Revue Biblique* 50 (1941) p. 239.

[6] Thus G. A. Downey, *A History of Antioch in Syria from Seleucus to the Arab Conquest* (Princeton 1961) p. 97 n. 52 'not clear'; Walbank in *Scripta Classica Israelica* 15 (1996) p. 126 n. 44 'A possibility, but no more'.

exceptional in the emphasis which was placed on the participation of the army in the festivities. The difference between this parade and that which Ptolemy Philadelphus held in Alexandria is the order in which the army participated. As Tarn puts it [7]

> 'The main feature of the show at Daphne was the review of Antiochus' strong army; it recalls the similar review of his army by Ptolemy II, but while at Alexandria the march past of the army came last, at Daphne it came first of everything.'

The units of the Seleucid Army that took part in the Daphne Parade are enumerated by Polybius, constituting one of the few descriptions of the Seleucid army which has passed down to modern times (See Figure 20). It is safe to assume that the units that paraded at Daphne did not comprise the Seleucid Army in its entirety, since garrisons must have been maintained in at least some of the more rebellious provinces. On the other hand these garrisons would, in the main, be furnished by units of mercenaries. It is reasonable to suppose that all the significant units of the Seleucid field army participated in the parade.

The most interesting feature of this passage, as far as the present work is concerned, is that it informs us that, marching at the very head of the parade, a substantial portion of the Seleucid army had already been reformed along Roman lines. Polybius (30. 25. 3) starts his description of the Parade as follows:

> καθηγοῦντό τινες Ῥωμαϊκὸν ἔχοντες καθοπλισμὸν ἐν θώραξιν ἁλυσιδωτοῖς, ἄνδρες ἀκμάζοντες ταῖς ἡλικίαις πεντακισχίλιοι·

> Heading were some men equipped in the Roman manner in mail cuirasses, five thousand men in the prime of life.

The mail cuirasses are clearly worn in imitation of the Roman *loricae hamatae*.[8] Launey notes that this is the first time that troops armed in the Roman manner had been recorded as appearing in the ranks of any Hellenistic army. It is surely superfluous to note, he added, that these troops are not ethnic Romans.[9] Given that they are described as men 'in the prime of life' they are likely to young men recruited from the Graeco-Macedonian population of the Seleucid Empire and should there-

---

[7] W. W. Tarn, *The Greeks in Bactria and India2* (Cambridge 1951) p. 194. The contrast with the Parade at Alexandria was first noted by Fritz Caspari, 'Studien zu dem Kallixeinos Fragment Athenaios 5, 197c - 203b' *Hermes* 68 (1933) p. 407 'nur daß diesmal die δυνάμεις den Anfang und nicht den Schluß bilden'.

[8] So Walbank, *Commentary* III p. 449.

[9] Launey I p. 609 'mais, soldats ou figurants, ce ne sont point, il est presque superflu de le dire, des Romaines'. Cf. E. Bikerman, *Institutions des Séleucides* (Paris 1938) p. 57 'C'était une des tentatives de ce roi d'imiter les institutions de la République victorieuse'.

Infantry.		
	5,000	'Roman' infantry.
	5,000	Mysians.
	3,000	Cilician light infantry.
	3,000	Thracians.
	5,000	Galatians.
(The 'Syro-Macedonian' Phalanx).		
	20,000	Macedonians.
	5,000	*Chalkaspides.*
		?*Argyraspides.*
(*Infantry*)	?46,000 +	(TOTAL)
Cavalry.		
	1,000	Nisaian Cavalry.
	3,000	Citizen Cavalry.
	1,000	Companion Cavalry.
	1,000	Friends.
	1,000	'Picked' Cavalry.
	1,000	The *Agema.*
	1,500	Cataphracts.
	9,500	(TOTAL)
Chariots and Elephants.		
	100	Six-horse Chariots.
	40	Four-horse Chariots.
	1	Four-Elephant Chariot.
	1	Two-Elephant Chariot.
	36	Elephants.

Figure 20. The Seleucid Army at the Daphne Parade, 166 BC.

fore be considered an element of the Seleucid 'Macedonian' phalanx. According to Bar-Kochva these 'young men in the prime of life' (ἄνδρες ἀκμάζοντες ταῖς ἡλικίαις) are an element of the Macedonian guard of the Seleucid army.

> 'The training of the 'Roman' contingent in the new style and their classification as picked troops suggests that this contingent could not have consisted of ordinary settlers called up in time of emergency, but was a permanent Guard.'[10]

It is Bar-Kochva's belief that the *Argyraspides*, or 'Silvershields' regiment, originally constituted the Guard of the Seleucid infantry. Under normal circumstances it was the only infantry regiment apart from the mercenary garrisons to be permanently embodied, the other regiments of the field army being mobilized only in time of war. He points out that the strength of the *Argyraspides* was normally 10,000 and noting that it numbered only 5,000 at the Daphne Parade, suggests that the balance was made up by the 5,000 infantry 'armed in the Roman manner'. He therefore believes that the 'Roman-armed' infantry also belongs to the Guard. In other words they are troops who would in prior times have been found serving in the *Argyraspides*.[11]

Argyraspides, or 'Silvershields', is a title which was first given to the élite Hypaspist infantry regiment of the Macedonian Army as a nick-name in 327. In this year the regiment was issued with silver shields as part of a number of changes in the army which Alexander the Great instituted prior to his invasion of India.[12] The nick-name subsequently became the official title of the regiment. Upon the collapse of Alexander's empire the regiment eventually passed into the service of Seleucus I.[13] It later became the élite regiment of the Seleucid infantry.[14]

The strength of the *Argyraspides* regiment at Daphne is not given. At the battle of Raphia in 217 Polybius (5.79.4) tells us that Theodotos the Aetolian commanded 10,000 men picked from the whole kingdom and armed in the Macedonian manner, most of them *Argyraspides*. It is probably this regiment which Polybius

---

[10] Bezalel Bar-Kochva, *The Seleucid Army* p. 60.

[11] Bar-Kochva, *Seleucid Army* p. 60; cf. G.T. Griffith, *The Mercenaries of the Hellenistic World* (Cambridge 1935) p. 146 'if the Roman-armed infantry are "Macedonians"'.

[12] Justin 12. 7. 5; cf. Curt. 8. 5. 4; Augustin Spendel, *Untersuchungen zum Heerwesen der Diadochen* (Breslau/Wroclaw 1915) pp. 43-53; W. W. Tarn, *Alexander the Great II. Sources and Studies* (Cambridge 1948) pp. 151 sq.; R. A. Lock, 'The origins of the Argyraspids' *Historia* 26 (1977) pp. 373-8; E. M. Hanson, 'Alexander's hypaspists and the argyraspides' *Historia* 30 (1981) pp. 117-20; A. B. Bosworth, *Conquest and Empire. The Reign of Alexander the Great* (Cambridge 1988) p. 270; N. G. L. Hammond, *Alexander the Great. King, Commander and Statesman*^{2} (Bristol 1989) p. 222.

[13] Plut., *Vit. Eum.* 16.1 - 19. 2; cf. Diod. 19. 43. 9.

[14] Edwyn Robert Bevan, *The House of Seleucus* (London 1902) II p. 285-6; Bar-Kochva, *The Seleucid Army* p. 58.

(5.85.10) later calls 'the picked men (*epilektoi*) of the Syrians'.[15] The implication of this passage is that at Raphia the regiment of *Argyraspides* formed part of the Macedonian phalanx, and so was composed of mobilized 'Macedonians'. The precise status of the individuals from which this élite regiment was formed is, however, unknown. Bar-Kochva has suggested that the *Argyraspides* was a permanently embodied 'guard' unit formed from the sons of military settlers.[16] This seems an acceptable hypothesis, providing that one substitutes the concept of 'Macedonian' for that of 'military settler'.

It is tempting to speculate that, while the Macedonian phalanx in its entirety was only mobilized in times of war, an élite regiment, called *Argyraspides*, was maintained on a permanent basis from those Macedonians who, having completed their obligatory military training, subsequently volunteered for continued service. The *Argyraspides* would, then, comprise the 'regular' component of the Macedonian phalanx.

Section 2 c

The Seleucid Phalanx at the Daphne Parade

The text of Polybius is unsatisfactory at the point where it comes to describe the Macedonian phalanx participating in the Daphne Parade. It runs as follows:

> τούτοις ἐπέβαλλον Μακεδόνες δισμύριοι καὶ χαλκάσπιδες πεντακισχίλιοι, ἄλλοι δὲ ἀργυράσπιδες,
>
> After these came 20,000 Macedonians and 5,000 *chalkaspides* (Bronzeshields) others *argyraspides* (Silvershields),

Griffith took the text as it stood, and translated it as '20,000 Macedonians, 5000 Bronze Shields, and some Silver Shields'.[17] Assuming that the original text written by Polybius was not itself grammatically defective, it is reasonable to assume that something has dropped out of the text. This fragment of Polybius is derived from the *Diepnosophistai* of Athenaeus which is, in turn, only preserved in a single manuscript, so we have no mechanism for restoring the lacuna through the normal methods of textual analysis. This has led to some rather free emendations of the

---

[15] Walbank, *Commentary* I p. 615 *ad loc.*.
[16] *Seleucid Army* pp. 59-62.
[17] G. T. Griffith, *The Mercenaries of the Hellenistic World* (Cambridge 1935) p. 146.

text, such as that of Kaibel, which is used for the Loeb translation of Athenaeus and Polybius, and which runs as follows.

> τούτοις ἐπέβαλλον Μακεδόνες δισμύριοι, <χρυσάσπιδες μὲν μύριοι> καὶ χαλκάσπιδες πεντακισχίλιοι, <οἱ δὲ> ἄλλοι ἀργυράσπιδες,
>
> After these came 20,000 Macedonians, 10,000 *chrysaspides* (Goldenshields) and 5,000 *chalkaspides* (Bronzeshields) the others being *argyraspides* (Silvershields),

Walbank rejects the emendation, noting that, besides the insertion, it involved altering ἄλλοι δέ to <οἱ> δὲ ἄλλοι.[18] A passage in *1Macc.* (6. 39; repeated in Josephus *AJ* 12. 372) describing the Seleucid Army at the battle of Beth-Zacharia has often been quoted in support of Kaibel's emendation. This passage mentions that as the sun rose on their shields of bronze and gold, 'the mountains glittered from them and blazed like torches of fire'. Many commentators regard this passage as one of high rhetoric, possibly influenced by Biblical passages which also mention shields of bronze and gold.[19] While it is perfectly possible that some of the shields of the cavalry or infantry units of the Seleucid army may have had gilded or bronze components, the passage is, in itself, insufficient evidence to postulate the existence of an otherwise unattested regiment called the *chrysaspides*, or 'Goldenshields', in the Seleucid Army.

References to *chrysaspides* in Greek literature do exist, but are extremely rare. Pallas Athena and other deities or heroes are called 'golden-shielded' in a number of passages in classical literature.[20]. The only regiment of *chrysaspides* for which we do have proper historical evidence, is that formed, along with a regiment of *argyraspides*, by the Emperor Severus Alexander prior to his expedition against the Sasanian king Ardashir in AD 232.[21] These regiments were obviously formed in imitation of the army of Alexander's Macedonian namesake. It is unknown if Severus Alexander created his 'Goldenshields' regiment on the basis of any particular literary text or not. The literary sources dealing with Alexander the Great's army are copious, but a regiment of *chrysaspides* is nowhere mentioned. The *Onomasticon* of Pollux of Naucratis lists *chrysaspides* (1. 175) along with *argyraspides* and *pezhetairoi* as terms suitable for giving to Macedonian infantry

---

[18] Walbank, *Commentary* III p. 450. Kaibel's emendation is accepted by Bar-Kochva, *The Seleucid Army* p. 232 n. 8 and *Judas Maccabaeus* p. 417 n. 21, pp. 325-6 where the existence of a *chrysaspides* regiment is accepted.

[19] Eg. *1 Kings* 14.26-7; *2 Chron.* 12.9-10.

[20] Pind., *Isth.* 1.1; Eurip., *Phoen.* 1372; Bacchyl., *Dithyramb* 19. 11.

[21] *Historia Augusta, Sev. Al.* 50.5.

regiments. It is as probable that Severus Alexander named his regiment on the authority of Pollux as it is that he had any historical knowledge now lost to us.[22]

The authority of Pollux is, however, a dubious one. The aim of works such as his *Onomasticon*, composed in an age when the Greek language was subject to considerable change, was to gather together long lists of obscure words no longer in common usage, which the reader could use to impress people at dinner parties. The ancient sources were accordingly dredged, and any strange word ripped from its context would fit the part. Pollux was quite ruthless in his methods, which were designed to impress the Emperor Commodus (AD 176-192) sufficiently so as to secure his appointment to the chair of rhetoric in Athens subsidized by the Emperor himself.[23] Pollux may well have come across a reference to golden shields in the ancient literature. It is even possible that Pollux had come across the term in the passage in *Maccabees* alluded to above, but many other possibilities exist. While, as has already been mentioned, no extant literary source mentions a regiment of 'golden-shields' in Alexander's Army, a number of passages in Curtius refer to the gold and silver equipment given to the army in the later years of the reign. Thus 8. 5. 4 refers to cuirasses decorated with gold and silver, 8. 8. 16 refers to arms decorated with gold and silver, and 9. 3. 21 refers to 25,000 panoplies inlaid with gold and silver. It is possible that some lost source referred to the 'golden shields' carried by the army, and that the word *chrysaspides* has been misquoted by Pollux.

Whatever Pollux's source may have been, we cannot take this stray reference to *chrysaspides*, which is not even given in a Seleucid context, as evidence for the existence of a discrete regiment of *chrysaspides* in the Seleucid army. It is logical to assume that there was only one élite regiment in the Seleucid phalanx, and that this was the *argyraspides* regiment.

Kaibel's insertion of *chrysaspides* into the text must be rejected. It seems preferable to assume that the original text of Polybius stood very much as it stands now, mentioning 20,000 Macedonians, 5,000 *Chalkaspides* and an unstated number of *Argyraspides*. As has already been mentioned, Bar-Kochva suggests that

---

[22] cf. A. B. Bosworth in *Harvard Studies in Classical Philology* 81 (1977) p. 246 n. 121. Walbank (*Scripta Classica Israelica* 15 (1996) p. 127 n. 52), whilst continuing to reject the restoration of 'Goldenshields' in the Polybian text, noted that a reviewer had added two references to 'Goldenshields' to those he had previously noted. However, of these two references, Onasander 1.20 refers to 'silvered and gilded panoplies' in an abstract and rhetorical context, while Plutarch *Vit. Eum.* 14. 5 speaks of the sun glinting on golden weapons (χρυσῶν ὅπλων). In the latter case, as has recently been restated (J.F. Lazenby and D. Whitehead, 'The Myth of the Hoplite's Hoplon' *CQ* n.s. 46 (1996) pp. 27-33) the work *hoplon* cannot be interpreted as 'shield'. Neither of these passages (also included in the discussion by Bar-Kochva, *Judas Maccabeus* p. 326) can be interpreted as referring to a unit of 'Goldenshields'.

[23] C. P.Jones, *Culture and Society in Lucian* (Cambridge Mass., London 1986) pp. 104, 108, 153.

the strength of the *Argyraspides* at Daphne may have been 5,000.[24] Although his suggestion piles one speculation on top of another, it ultimately seems plausible. We might further speculate that the reason why Polybius' text has become corrupt is that it originally mentioned 20,000 Macedonians, 5,000 *Chalkaspides* and 5,000 *Argyraspides*, but that the second mention of the figure 5,000 has dropped out of the text.

Given that the received text is demonstrably corrupt, one cannot be sure whether the figure of 20,000 Macedonians is meant to be inclusive of the 5,000 *Chalkaspides* and the *Argyraspides* or exclusive of them. Taking the Greek of the passage at face value, and given the καί which occurs in the text between the Macedonians and *Chalkaspides*, the probability would seem to be that the Macedonian phalanx numbered 20,000 exclusive of the other regiments. Thus the total strength of the mobilized Syro-Macedonian phalanx in 165 would be 35,000 if we number the *argyraspides* at 5,000 and include the 5,000 men equipped 'in the Roman manner'.

An 'exclusive' interpretation is accepted by Griffith,[25] who suggests that this constitutes a significant increase over the statistics for Magnesia; and Tarn,[26] who takes the rise in numbers as conclusive proof for a considerable influx of Greek settlers in Syria between Magnesia and the Daphne Parade. On the other hand Launey,[27] while undecided on the total strength of the Macedonian phalanx at Daphne, suggests that the number of these 'Macedonians' may well have been swelled by orientals drafted into the ranks for the parade.

Bar-Kochva adopts a more sophisticated approach.[28] Although accepting an 'inclusive' interpretation, he suggests that the strength of the Seleucid phalanx had remained more or less stable. At Raphia the number of the Seleucid phalanx was 20,000 plus the 10,000 who were mostly *argyraspides* (Polyb. 5. 79. 4-5). It seems probable that some 6,000 Kyrrhestai were not present at the battle, having mutinied over pay some time earlier (Polyb. 5. 50. 8). They were presumably still in revolt (5. 57. 4). Those Macedonians settled in the west of the Empire were also not present, as Achaios had risen in revolt. Bar-Kochva suggests their strength would have been around 6,000, as this is the size of the force that Achaios used against Selge in 218 (Polyb. 5. 72. 3). Whether this last speculation is correct or not, if Bar-Kochva is correct about the Kyrrhestai the Seleucid phalanx recruited in Asia would have had a strength of 36,000. Bar-Kochva also notes that in addition to the 16,000 phalangites and probably 10,000 *argyraspides* mentioned at Magnesia, the potential size of the phalanx must have included some of the troops who had been en-

---

[24] Bar-Kochva, *The Seleucid Army* p. 60.

[25] G. T. Griffith, *Mercenaries of the Hellenistic World* (Cambridge 1935) p. 146.

[26] W. W. Tarn, *The Greeks in Bactria and India2* (Cambridge 1951) p. 186.

[27] Launey I p. 319; a possibility also accepted by F. W. Walbank, *The Hellenistic World* (London 1981) p. 130.

[28] Bar-Kochva, *The Seleucid Army* p. 41.

gaged in the earlier campaign in Greece, at least part of whom had been lost. Antiochus had about 10,000 Seleucid infantry at his disposal in Greece, at least some of which were *sarisophoroi* of the Macedonian phalanx (Livy 36. 15. 3; 36. 18. 2). This analysis of statistics concerning the Seleucid phalanx suggests that a figure of about 35,000 men for the Daphne Parade is in the right area.

We may finally turn to look at the *Chalkaspides*. Just as *argyraspides* was a regimental title specific to the Seleucid Army, and all references to *argyraspides* occur within Seleucid contexts, *Chalkaspides* is a regimental title specific to the Antigonid Army, and all references to *Chalkaspides* apart from this one occur in Antigonid contexts. A second exception proves the rule. Shortly before the Battle of Sellasia in 222 King Antigonos Doson had given a quantity of bronze Macedonian shields to the citizens of Megalopolis, one of the constituent cities of the Achaean League. The Megalopolitans were therefore able to send a force of 1,000 troops 'armed in the Macedonian manner' to participate in the battle (Polyb. 2. 65. 3). When the Social War broke out in 219 Philip V of Macedon force marched to the Peloponnese to help his Achaean allies together with a force including 3,000 *Chalkaspides*. When this regiment took the field their Eleian opponents at first took them to be Megalopolitans (Polyb. 4. 69. 4). As soon as the Eleians realized their mistake, they broke and fled.

What, then, is the reader to make of Polybius's reference to a regiment of *chalkaspides* participating in the Daphne Parade in 166? Two possible explanations come to mind. The Antigonid regiment of *Chalkaspides* had ceased to exist after the battle of Pydna in 168, and one explanation might be that Antiochus reformed the regiment within his army from exiles and fugitives fleeing from the defeated Macedonian army. It is reasonable to suppose that after the disaster at Pydna, there was an exodus of Macedonians who had previously made their living as professional soldiers in the Antigonid Army. Whether it is reasonable to suppose that this number would have amounted to 5,000 men, the size given for the Seleucid *Chalkaspides* regiment in 166, is another question. Thus the regiment would have been formed shortly before 166. An alternative explanation for the appearance of a the *Chalkaspides* at Daphne in 166 might be that this regiment represented an earlier attempt by Antiochus, or possibly by his more pro-Macedonian predecessor Seleucus IV, to reform the Seleucid Army, this time along Macedonian lines.

The Parade of Daphne is the last occasion that we hear of units of either *Argyraspides* or *Chalkaspides* in the Seleucid army. One can only assume that both regiments disappeared soon after 166 as the process of 'Romanization' of the Seleucid Army grew apace.

Section 2 d

The Strength of 5,000 and the Date of the Reform

The strength of 5,000 given to the body of Roman-armed infantry which participated in the Daphne Parade could be interpreted in two ways. In the first place, if we see the number of 5,000 as the permanently established strength of the formation, then this could be interpreted to suggest that Antiochus had created an experimental 'legion'.[29]

Polybius gives 4,200 as the standard size of the legion, mentioning that it could be expanded to 5,000 in times of exceptional danger (6.20.8; 2. 107. 10). Of the normal total, however, 1,200 were light infantry or *velites* and only 3,000 heavy infantry 'equipped in the Roman manner'. The statistics for legionary strengths have been gathered and discussed by Brunt.[30] During the Second Punic War legionary strengths in excess of 5,000 men are mentioned. During the course of describing the forces at the disposal of Rome in 225 BC on the eve of Hannibal's invasion of Italy, Polybius (2. 24. 3) gives a strength of 5,200 foot for each of the eight legions of Roman citizens commanded by the consuls and a strength of 4,200 for each of the two reserve legions stationed in Sicily and Tarentum (2. 24. 13). The reserve force of 20,000 Roman citizen infantry stationed in Rome itself probably comprised a further four legions of 5,000 (2. 24. 9).[31] An exceptionally large army of eight legions was mobilized for Cannae in 216, each consisting of 'about five thousand men' (3. 107. 9). By the time of the Third Macedonian War the strength of the legion sometimes rose to 6,000, and legionary strengths of 6,200 are reported of the army which the elder Scipio transported to Africa in 204 BC (Livy 29. 24. 14) and the legion Gaius Flaminius raised for service in Spain in 193 BC (Livy 35. 2. 4). We do not know how such a large legion would have been organized, but if we deduct 1,200 for the strength of the velites, we are left with a total of 5,000 for the rest of the legion. This might have been organized so that the 20 maniples of *hastati* and *principes* numbered 200 men each, and the 10 maniples of *triarii* may have numbered 100 each.[32] Thus the men 'equipped in the Roman manner' at the Daphne

---

[29] Cf. Mørkholm, *Antiochus* p. 39 '5000 strong or a little less than a legion, armed after the Roman pattern'.

[30] P. A. Brunt, *Italian Manpower 225 B.C. - A.D. 14* (Oxford 1971) Appendix 25 'The Size of the Pre-Marian Legion' pp. 671-676; cf. Walbank, *Commentary* I 702-3.

[31] Cf. Walbank, *Commentary* I p. 196.

[32] Thus Nick Sekunda and Angus McBride, *Republican Roman Army 200-104 BC* (London 1996) p. 36.

Parade perhaps numbered 5,000 because this was the usual number of heavily-armed men in a Roman legion.

This could be one interpretation. If this were the case, however, we would have to admit that the aims of Antiochus were limited to a piecemeal reform of the army and the creation of a single Seleucid 'legion'. Antiochus IV Epiphanes may have introduced reform into the Seleucid infantry along Roman lines at any point in his reign before this type of troop is first attested in the army in 166 at the Daphne Parade.

Nevertheless three dates suggest themselves as more probable than others. The earliest possible date is immediately after Antiochus' accession to the throne in 175. Antiochus had spent much of his earlier life as a hostage in Rome where, like Polybius later on, he became aware of the efficacy of Roman institutions. Although certainty is not possible, it is unlikely that Antiochus would have restricted his aims to the creation of a single 'legion', and that no further progress in infantry reform was made between 175 and 166.

Alternatively, the number 5,000 might be coincidental, and it might be that the figure represents a temporary stage in a process of complete re-training and re-arming of the Seleucid army. This process would necessarily have taken time, and the process had not yet had time to touch the whole of the Seleucid phalanx. Before the reforms of Antiochus Epiphanes they would have been trained in the weapon-handling and tactics of the Macedonian phalanx, but now they had been given Roman equipment and had been trained in manipular tactics instead. Thus these 5,000 men 'in the prime of life' represent a specific number of age-classes of the population of the Seleucid Empire who were liable for military training and who had already received Roman-style equipment and training. In other words that the number 5,000 represents a number of age-classes of the Seleucid male population subject to military service rather than an organizational structure.

The second date which suggests itself as being probable for the introduction of reform into the Seleucid infantry is during the military build-up which preceded the outbreak of the Sixth Syrian War in 170. Once again one would have to conclude either that Antiochus' aims had been limited to the creation of a single Roman-Armed 'legion', or that very little progress had been made in the process of reform of the Seleucid infantry before 166.

The last of the three probable dates for the introduction of the reform seems the most suitable to the present author. This date would be after the unsuccessful conclusion of the Sixth Syrian War, when, following the defeat of the Macedonian Army at Pydna on 22 June 168, a Roman embassy headed by Popilius Laenas ordered Antiochus to leave Egypt. The whole Hellenistic world suffered an enormous culture-shock when it heard of the overthrow of the Macedonian military system by the Romans, first at Kynoskephalai in 197 and then at Pydna in 168. The astonishment which was felt throughout the Hellenistic world is echoed in the words with which Polybius concludes his comparison of the Macedonian phalanx and the

Roman manipular system. It is surely significant that this discussion comes in Polybius' text immediately after his account of the Roman victory at Kynoskephalai. He thought it necessary to speak on this subject at some length, he tells us,

> περὶ μὲν <οὖν> τούτων ἀναγκαῖον ἡγησάμεν εἶναι τὸ διὰ πλειόνων ποιήσασθαι μνήμην διὰ τὸ καὶ παρ' αὐτὸν τὸν καιρὸν πολλοὺς τῶν Ἑλλήνων διαλαμβάνειν, ὅτε Μακεδόνες ἡττήθησαν, ἀπίστῳ τὸ γεγονὸς ἐοικέναι, καὶ μετὰ ταῦτα πολλοὺς διαπορήσειν διὰ τί καὶ πῶς λείπεται τὸ σύνταγμα τῆς φάλαγγος ὑπὸ τοῦ Ῥωμαίων καθοπλισμοῦ.
>
> I thought it necessary to speak on this subject because many Greeks on the actual occasions when the Macedonians suffered defeat considered the event as almost incredible, and many will continue to wonder why and how the phalanx comes to be conquered by troops armed in the Roman fashion.[33]

This passage not only illustrates the shock to the system which the military defeats at Kynoskephalai and Pydna constituted to Greek and Macedonian attitudes, it also conveys some of the disbelief that all civilized cultures experience when they suffer military defeat at the hands of 'barbarians'. The shock was particularly enhanced because the Romans placed principal trust in hand-to-hand combat and the sword: literally in 'cold steel'. The blade of the Spanish sword used by the Romans was very strong and firm and so formed an excellent thrusting weapon, and was also sharpened on both edges for cutting.[34] The method by which this Spanish steel was produced is preserved in a fragment of Poseidonius describing the weapons of the Celtiberians. Iron plates would be buried and only excavated when the rust had eaten away the softest part of the metal. The remainder would then be worked into swords of the most unyielding metal.[35]

The Spanish sword had a tremendous impact on the Macedonians. Livy describes the dismay that swept through the Macedonian army when they saw the damage the Spanish sword had inflicted on the bodies of the dead during a skirmish in the early stages of the Second Macedonian War. King Philip had hoped to stir the Macedonian army into eagerness to avenge the fallen by displaying their bodies, but[36]

---

[33] Polyb. 18.32.13, trans.Paton, Loeb.

[34] Polyb. 6. 23. 6-7. Polybius states that the Spanish sword was adopted during the Second Punic War, but its adoption may have taken place somewhat earlier (Walbank, *Commentary* I p. 704, III p. 754).

[35] Diod. 5. 33.

[36] Livy 31. 34. 3-5 trans. Sage, Loeb.

> Quod promptiores ad subeundam omnem dimicationem videbatur factu- rum, id metum pigritiamque incussit; nam qui hastis sagittisque et rara lanceis facta vulnera vidissent, eum Graecis Illyrisque pugnare adsueti, postquam gladio Hispaniensi detruncata corpora, bracchiis eum humero abscisis, aut tota cervice desecta divisa a corpore capita patentiaque viscera et foeditatem aliam vulnera viderunt, adversus quae tela qu- osque viros pugnandum foret, pavidi vulgo cernebant.

> What he thought would make them more ready to enter any conflict caused, instead, reluctance and fear; for men who had seen the wounds dealt by javelins and arrows and occasionally by lances, since they were used to fighting with Greeks and Illyrians, when they had seen bodies chopped to pieces by the Spanish sword, arms torn away, shoulders and all, or heads separated from bodies, with the necks completely severed, or vitals laid open, and the other fearful wounds, realized in a general panic with what weapons and what men they had to fight.

The Macedonians had once been the barbarians from the north, but they had be- come too civilized. A new and much more fearful barbarian had arrived from the west.

It seems possible to argue that Antiochus only embarked upon the reform of the Seleucid infantry after his withdrawal from Egypt in 168. This reform would presumably have been carried out in preparation for his Eastern Campaign, upon which he embarked after the Daphne Parade. The fact that the 5,000 march at the head of the whole army is surely of the utmost importance. Perhaps by this act Antiochus aimed to show his intention of reforming the whole of the Seleucid heavy infantry along Roman lines.

Section 2 e

The Seleucid Army at Beth-Zacharia

The next glimpse of the Seleucid army in the literary sources comes in a description of the battle of Beth-Zacharia in 162 BC. Both *1 Maccabees* (6. 30) and Josephus (*AJ* 12. 366) give the strength of the Seleucid forces as about 100,000 infantry, 20,000 cavalry and 32 elephants. In *2 Macc.* 13. 2 we are told that the Seleucid force consisted of Greek troops numbering 110,000 infantry, 5,300 cavalry, 22 elephants, and 300 scythed chariots, and elsewhere Josephus contradicts him- self too, for in *BJ* 1. 41 the numbers are given as 50,000 infantry, 5,000 cavalry and

80 elephants. It is difficult to chose between these discrepant accounts of the strength of the Seleucid army present at the battle. Abrahams accepts the number of 32 elephants as being the most probable, being closest to the figure of 36 single fighting elephants, excluding chariot elephants, which Antiochus IV paraded at Daphne.[37] Both *1Maccabees* (6. 35) and Josephus (*AJ* 12. 371) tell us that the elephants were distributed among the phalanxes. A thousand infantry were stationed with each of them, and 500 picked horsemen were also assigned to each animal. Needless to say, we should interpret the word phalanx in these texts in the ancient sense as meaning 'a group of heavily-armed troops in formation' rather than in the modern sense as meaning 'a group of armoured pikemen in formation'. Taking these two pieces of information together at face value, it would seem that the Seleucid heavy infantry present at the battle numbered 32,000 men. This figure would be very close to the totals given for the Seleucid heavy infantry above, arrived at by the calculations of Bar-Kochva. The tactical plan is strikingly reminiscent of the 'articulating phalanx' formed by the Seleucid army at Magnesia, where the phalanx was divided into ten blocks each separated by two elephants, a total of 22 elephants in all.

More importantly from our point of view, *1 Macc.* 6. 35 also tells us that the infantry of the phalanx stationed beside the elephants wore mail cuirasses and bronze helmets (τεθωρακισμένους ἐν ἁλυσιδωτοῖς καὶ περικεφαλαῖαι χαλκαῖ ἐπὶ τῶν κεφαλῶν αὐτῶν). The implication of the passage is that the whole of the Seleucid phalanx present at the battle, whatever its total number, was now armed in the Roman manner. Bar-Kochva does not interpret the passage in this way. He assumes that there remained only one unit of Roman-armed infantry in the Seleucid army, and that the majority of the Seleucid infantry participating in the battle were still equipped as phalangites. He assumes that the eyewitness source of *1 Maccabees* was evidently carried away by the sight of the one contingent armed in the Roman style 'because they were in the forward formations which entered the pass and so caught the spectator's eye, and also because of the particular biblical association evoked'.[38] It would be more reasonable to read the passage as it stands and to assume that the whole of the phalanx is being described.

The battle at Beth-Zacharia involved an advance through a defile which was only 100 metres wide at its narrowest point.[39] Each of the infantry units accompanying the elephants would have had to contract its frontage to a width sufficiently small to pass through this defile. Considerable skill and a high level of individual training would have been required to deploy each of the army detachments into line, then into column to negotiate the defile, then back into line again. This fact has been noted by Bar-Kochva. Curiously he uses this very flexibility as an argument that

---

[37] Israel Abrahams, *Campaigns in Palestine from Alexander the Great* (London 1927) p. 30.
[38] Bar-Kochva, *Seleucid Army* p. 55; *Judas Maccabaeus* p. 315.
[39] Bar-Kochva, *Judas Maccabaeus* p. 330.

the Seleucid Army was still predominantly a phalangite force, and that even the Roman-armed infantry unit had not adopted the more flexible Roman infantry organization and tactics, but only Roman equipment.[40] In the view of the present author the flexibility of Seleucid formations at Beth-Zacharia is an indication that the Seleucid army had adopted Roman tactics as well as equipment. The great weakness of the traditional Macedonian pike-phalanx was its inflexibility.

It could be argued that between the Parade of Daphne in 166 and the battle of Beth-Zacharia in 162 the Seleucid phalanx had been entirely retrained in the use of Roman weapons and tactics. This could have taken place during the course of the Eastern Campaign of Antiochus IV.

Section 2 f

Higher Organization of the Seleucid Infantry

In Appendix B it is suggested that the organization of the heavy infantry, and the names given to its various formations in Asclepiodotus and the other authors of the Greek *Taktika* go back to an original manual composed by Poseidonius of Apameia, and that it is possible that some of the names given to various formations are based on late Seleucid military practice. As we have seen above, it is difficult to trace the word *syntagma* being used of a sub-unit in either the Macedonian army under Alexander, or in its Antigonid or Ptolemaic successors. It is therefore possible, though no more than that, that the new Seleucid maniple was called a *syntagma*. As has already been mentioned, the normal Greek word used for a maniple is *sêmeia* in the papyri and *sêmaia* in Polybius. This word is attested for the Seleucid army. Polybius (10. 49. 7) mentions that in 208 during his Bactrian Campaign Antiochus III ordered his cavalry to form up in *sêmaiai* and *oulamoi*. It is possible, therefore, that the normal Greek term already in use to designate the infantry maniple may not have been adopted by the Seleucid army because the word *sêmaia* was already in use for a cavalry formation. It is surely not necessary to stress the highly speculative nature of this suggestion.

In the course of his description of the various formations to be adopted by the infantry *syntagmata*, Asclepiodotus mentions (11.7) a loose formation, when the *syntagmata,* each a square block of 16 by 16 men, are arranged 'by parallelograms' (*kata parallêlogramma*) with only the corners touching. This formation is illustrated by a line-drawing in the text of Asclepiodotus.[41] This is one of a number

---

[40] Bar-Kochva, *Judas Maccabaeus* 126-9.

[41] A line drawing of this diagram also appears in *Aeneas Tacticus, Asclepiodotus, Onasander*, with

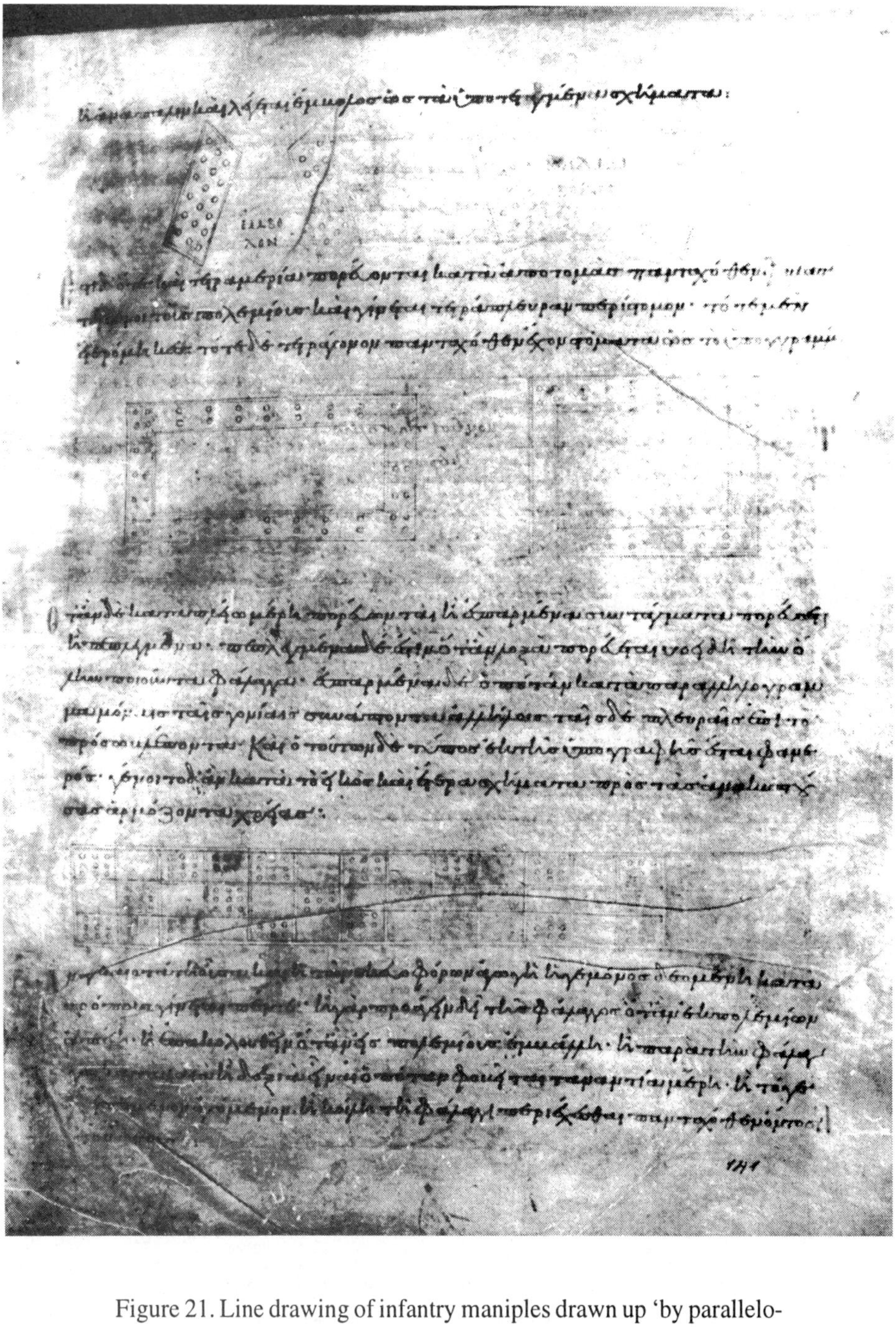

Figure 21. Line drawing of infantry maniples drawn up 'by parallelograms' (at bottom) in the manuscript of Asclepiodotus (photograph: Bibliotheca Mediceo-Laurenzia, Florence).

of such drawings in the manuscript which are referred to by the author. Clearly the original draft of the manuscript was illustrated with line drawings. Asclepiodotus is presumably describing a Late Hellenistic version of the Roman 'chequer' formation.[42] The prototype manuscript of Asclepiodotus (*Laurentianus* LV-4), which is now in the Bibliotheca Mediceo-Laurenziana in Florence, is a carefully copied document issued by the *scriptorium* of the Byzantine Emperor Constantine VII Porphyrogenetos in the middle of the tenth century AD.[43]

In Asclepiodotus two *syntagmata* make up a *pentakosiarchia* of 512 men under a pentakosiarch, and two *pentakosiarchiai* made up a *chilarchia* commanded by a *chiliarchês*. Chiliarch is, it should be noted, the term used by Polybius (6. 19. 7) for the Roman military tribune, and it occurs in Imperial inscriptions as a term for *tribunum militum*.[44] One suspects that the *pentakosiarchia* at least is fictitious, added to the mathematical constructions of the text by Asclepiodotus or Poseidonius previously on the basis of early Hellenistic military practice. For Roman tactics, including the *acies triplex*, to work, maniples should not simply double. One would rather expect them to multiply by three or six. One likewise suspects that the strength of the century and maniple given by Asclepiodotus have been exaggerated for mathematical purposes. Once again, it is possible that the terminology preserved in the *Taktika* and originally derived from Poseidonius preserves late Seleucid military practice. It is possible, therefore, that a number of maniples, perhaps six as in the contemporary Ptolemaic army, may have been grouped in formations called *chilirchiai*. We have seen that the ranks of pentakosiarch and chiliarch seem not to survive the reform of the 160's in the Ptolemaic army, however the *chiliarchia* may have existed in the late Seleucid army as is suggested by two pieces of literary evidence.

A fragment from the thirty-fourth book of Poseidonius' *Histories*, preserved in Athenaeus (4. 153b) informs us that the Seleucid general Herakleon, a contemporary of Poseidonius as is noted in Appendix B, made his soldiers take their meals by thousands, seated on the ground in the open air. Dinner consisted of a large loaf with some meat, and wine tempered with water to drink. 'Sword bearers' (μαχαιροφόροι) used to serve the troops, and there was an orderly silence. These 'sword bearers' seem to be servants attached to the army. Perhaps they can be

---

an English translation by members of the Ilinnois Greek Club (Loeb ed., Cambridge Mass. 1923) p. 322.

[42] Noted by Everett L.Wheeler, 'The Legion as Phalanx' *Chiron* 9 (1979) pp. 303-18 esp. p. 306 n. 14.

[43] Lucien Poznanski, *Asclépiodote Traité de Tactique* (Budé, 1992) p. xvi.

[44] Thus in an inscription from Delphi dating from the first decade of the second century AD a *praefectus cohortis* is termed a ἔπαρχος σρείρης and a *tribunus militum* is a χειλίαρχος (Gaston Colin, *Fouilles de Delphes Tome III Épigraphie, Fascicule IV Inscriptions de la Terrasse du Temple et de la Région nord du Sanctuaire Nos. 1-86* (1930) pp. 72-77 no. 48). Cf. Hugh J. Mason, *Greek Terms for Roman Institutions* (American Studies in Papyrology Vol. 13, Toronto 1974) p. 164.

compared to the Egyptian *machairophoroi* mentioned in contemporary Ptolemaic papyri and inscriptions such as the 'Jouget' and 'Roeder' stelai from Hermopolis. As has been noted in the previous section both Josephus and *1Maccabees* tell us that at the battle of Beth-Zacharia in 162 BC the Seleucid Army was divided into 32 tactical units, each grouped around one of the 32 elephants of the army, and each consisting of 1,000 infantry and 500 cavalry.

Asclepiodotus (2. 10) tells us that two chiliarchies made up what was formerly called a 'wing' (*keras*) or 'component' (*telos*),[45] commanded by a *telarchês*, but later a *merarchia* commanded by a *merarchês*. The rank of *meridarchês* is attested in the Seleucid Empire in the second century, though only as an administrative post in charge of a 'division' (*meris*) of a satrapy.[46] It is possible that the titles *merarchia* and *merarchês* had existed at some point in the Seleucid army. If we compare the organization of the Seleucid phalanx of 16,000 men at the battle of Magnesia we find it divided into ten *merê,* or 'divisions' of 1,600 (Appian, *Syr.* 32). It is possible to regard *merarchia* as a variant for *meros* and *merarchês* could have been the name of its commander. This would be extremely speculative. Rather than speculate that a term used in the Seleucid army before the reforms of the 160's was still in use in the late Seleucid army, it is perhaps more probable that an obsolete term has been used to dress with a name a further fictitious mathematical formation.

Further up the organizational chain Asclepiodotus (2. 10) has two *merarchiai* make up a body of 4,096 men, which is 'even now' called a *phalangarchia* or a 'half-wing', commanded in former times by a *stratêgos* but now by a *phalangarchês*, two of which made up a *diphalangia* or a 'wing' commanded by a *kêrarchês* or 'wing-commander'. Two of these made up the 16,384 strong phalanx of heavy infantry. We have already noted that the figure of 16,000 is given as the strength of the Seleucid phalanx at the battle of Magnesia (exclusive of the *Argyraspides*).[47]

At the battle of Raphia in 217 the Seleucid phalanx is stated (Polyb. 5. 79. 5) to have been under the command of Nicarchus and of Theodotus 'Hemiolios'. This need not necessarily imply any formal division of the phalanx into two tactical units called 'wings', but this would be the natural interpretation of the passage. The centre of the Seleucid line at Magnesia, which consisted of the Seleucid phalanx

---

[45] In later Greek texts the term *telos* is regularly used as one of the terms for a legion (Hugh J. Mason, *Greek Terms for Roman Institutions* (American Studies in Papyrology Vol. 13, Toronto 1974) p. 164).

[46] *1 Macc.* 10. 65; Joseph., *AJ* 12. 261, 264, 15. 216. On these passages see Bezalel Bar-Kochva, *Judas Maccabeus* (Cambridge 1989) p. 202. On the office see Hermann Bengtson, *Die Strategie in der Hellenistischen Zeit* 2 (Munich 1944) p. 27, 172 but cf. W. W. Tarn, *The Greeks in Bactria and India*^{2} (Cambridge 1951) p. 242 n. 6 'The usual view is that the meridarch governed some subdivision of the satrapy, but the difficulties are great'.

[47] Livy 37.40.1; Appian, *Syr.* 32.

broken up into ten blocks with two elephants in between each gap, was commanded by Minnion, Zeuxis and Philippos 'the elephantarch' (Livy 37. 41. 1). One way in which the information given by Livy could be interpreted is to understand that Philippos would have been in at least notional command of the elephants, dispersed though they were, and therefore that Minnion and Zeuxis each commanded one wing of the phalanx. Finally it should be noted that *1 Maccabees* mentions that King Demetrius I sent 'the right wing' (τὸ δέξιον κέρας) of the Seleucid army into Judaea under the command of Bacchides and Alkimos in 160 BC.[48] The context implies that 'the right wing' was, indeed, a standing division of the army, a fact which was noted by Bevan.[49]

There is furthermore a little possible evidence for the organization of Seleucid armies into 'half-wings'. Polybius (5. 53. 6) mentions that during the suppression of the revolt of Molon in 220 Antiochus the Great took command of the right wing himself and entrusted command of the left wing to Hermeias and Zeuxis. It must be stated, however, that our picture of the disposition of the Seleucid army during this battle is very sketchy. Polybius does not seem to be thinking of the heavy infantry being divided into two wings, but rather the whole army. He previously states (5. 53. 4) that the Companion cavalry were stationed on the left wing next to the phalanx. Consequently we do not really know whether Hermeias and Zeuxis divided command of just the Companion cavalry between them, or of the Companion cavalry and the phalanx or half the phalanx. We have already noted that *1 Maccabees* mentions that King Demetrius I sent 'the right wing' of the Seleucid army into Judaea under the command of Bacchides and Alkimos in 160 BC. If we could be sure that this was the right wing of the heavy infantry, we then might be able to regard Bacchides and Alkimos as 'half-wing' commanders, but this assumption is likewise uncertain.

In summary, then, the formal division of the phalanx into a right and left wing and into half-wings may to some degree represent late Hellenistic, and specifically Seleucid, military practice rather than mere Poseidonian or Asclepiodotan caprice. If the Seleucid army at Beth-Zacharia in 162 was divided into 32 *chiliarchiai*, as is implied by the account preserved in Josephus and *1 Maccabees*, then eight of these could have made up a 'half-wing'.

---

[48] *1 Macc.* 9.1; cf. Joseph. *Ant. Jud.* 12.426-9.

[49] Edwyn Robert Bevan, *The House of Seleucus II* (London 1902) p. 284 n. 7; contra Bar-Kochva, *Judas Maccabaeus* p. 382.

Figure 22. Cast of Terracotta in Pennsylvania University Museum, accession number CBS 1955. 102 (photograph: Pennsylvania University Museum).

Figure 23. Hemidrachm struck by Antiochus IV in Antioch about 170 (photograph: Courtesy of the British Museum).

Section 2 g

The Nippur Terracotta

As the archaeology of the Seleucid Empire is known much less completely than that of Ptolemaic Egypt, it would not be surprising if no archaeological evidence at all had survived demonstrating that the equipment of the Seleucid infantry had changed. Only one terracotta has survived which possibly represents a Seleucid infantryman equipped in the Roman manner.

Figure 22 shows a cast, now in the Pennsylvania University Museum, of a Hellenistic terracotta figurine recovered during excavations of the University of Pennsylvania in the ancient Mesopotamian city of Nippur.[50] The original terracotta is in Istanbul. It is some 55 mm. high, and, though without an archaeological date, it clearly comes from the Hellenistic period. Due to its poor state of preservation the details are difficult to delineate and interpret. The subject may represent a Seleucid Roman-armed infantrymen rather than a traditional Greek *thureophoros* infantryman. Despite its crudity, the shoulder-guard of a mail cuirass can perhaps be made out below the right shoulder. Furthermore, the sword, on which the sword-hand

[50] Leon Legrain, *Terra-Cottas from Nippur* (*University of Pennsylvania, The University Museum, Publications of the Babylonian Section Vol. xvi*, Philadelphia 1930) p. 22, pl. xxv no. 131.

rests, is worn on the right hip in the Roman manner, rather than being slung under the left armpit in the Greek manner. This would be an extremely rare pose for a traditionally-armed Greek warrior. In fact the original publisher Legrain thought he was carrying 'a dart' in his right hand. The irregularity in the centre of shield could perhaps be a simple metal boss, or possibly the remains of one of the traditional Seleucid military badges, a 'Medusa Rondanini' head lying on a bed of scales.

The so-called 'Medusa Rondanini' is a Hellenistic variant of the gorgoneion symbol, where the features have become softened and have been given a sense of pathos. Two small wings are shown on the forehead and two snakes are knotted below the chin. Frequently, but not necessarily, the 'Medusa Rondanini' is shown lying on a bed of scales. The use of the emblem of the 'Medusa Rondanini' head as a Seleucid symbol has been dealt with in an article by Peter Callaghan.[51] Pausanias (5. 12. 4; cf. 1. 21. 3) tells us that one king Antiochus donated a golden head of Medusa the Gorgon with an aegis placed round it which was displayed on the south wall of the Athenian Acropolis facing the theatre.

Dispute has ensued as to which Antiochus made the gift. Some scholars have held that the gift was made by Antiochus III while others maintain that it was a donation of Antiochus IV. Callaghan notes that the 'Medusa Rondanini' appears as an emblem in Cretan pottery in the last two decades of the third century, and he suggests that the ultimate inspiration for the introduction of the emblem into the Cretan iconographic repertoire was the golden aegis set up in Athens by the Seleucid monarch, who in this case must have been Antiochus III. He notes, however, that the gorgoneion alone, without the bed of scales, is a well-established Seleucid symbol.

An interesting contemporary example of the Seleucid use of the 'Medusa Rondanini' on a bed of scales comes on the reverse of a coin struck in Antioch around 170 (Figure 23). The obverse shows the radiate head of Antiochus IV, while on the reverse there is a representation of Medusa's head on an aegis. Babelon was the first to connect this coin with the Seleucid gorgoneion from the Athenian acropolis, but it has been Mørkholm who has remarked most comprehensively on the unusual features shown on the coin.

> 'In contrast to all other representations of this subject on Greek coins, where the head of Medusa tends to become the dominating feature, the interest of the artist has here been concentrated on the aegis itself, the head being subordinated to the whole. Furthermore the aegis is not the usual pliable goatskin, normally drawn with a circular outline, but gives the impression of being made of a hard and stiff material such as wood

[51] P. Callaghan, 'The Medusa Rondanini and Antiochus III' *BSA* 76 (1981) pp. 59-70. For the dispute as to which Antiochus made the dedication see Habicht in *Chiron* 19 (1989) pp. 11-4.

or metal carved into a peculiar shape. On some of the dies ... there are indications that the aegis was to be fastened with six nails, one at each corner and one in the middle of each vertical side, to a wall or similar structure. On other dies the aegis appears to be fastened only at the four corners.'[52]

Even if this coin does show the golden Medusa and aegis donated on the Athenian Acropolis, it cannot be taken as proof that Antiochus IV Epiphanes made the donation. The issue of the coin may have been intended to recall the gift made by Epiphanes' father Antiochus III. Either way, these coins do show the importance of the Medusa Rondanini in Seleucid iconography, and it is perfectly possible that the Medusa Rondanini was used as a shield device by troops in the army of Antiochus IV.

Figure 24. Helmet shown on a coin of Tryphon in the Collection Luynes (3376) Cabinet des Médailles, Bibliothèque Nationale, Paris (photograph: Bibliothèque Nationale).[53]

[52] Otto Mørkholm, *Studies in the Coinage of Antiochus IV of Syria* (Historisk-filosofiske Meddelelser udgivet af Det Kogelige Danske Videnskabernes Selskab, Bind 40 nr. 3, Copenhagen 1963) pp. 23; cf. Mørkholm, *Antiochus* p. 59. See also Ernest Babelon, *Catalogue des Monnaies grecques de la Bibliothèque nationale, Les Rois de Syrie, d'Arménie et de Commagène* (Paris 1890) p. xcii and for other examples and references Arthur Houghton, *Coins of the Seleucid Empire from the Collection of Arthur Houghton* (New York 1983) no. 102-3. A similar gorgoneion is shown on later coins of Alexander I Balas (Houghton, *op. cit.* nos. 207-8).

[53] Published by Ernest Babelon, *Catalogue des Monnaies grecques de la Bibliothèque nationale, Les Rois de Syrie, d'Arménie et de Commagène* (Paris 1890) no. 1043.

Figure 25. Coin of Timarchus in the British Museum, Department of Coins and Medals 1913-5-12-1 (photograph: Courtesy of the Trustees of the British Museum).

Figure 26. Coin struck by Alexander I Balas in Antioch, in the British Museum, Department of Coins and Medals 1931-4-6-264 (photograph: Courtesy of the Trustees of the British Museum).

As often happens, the head of the terracotta is missing. If we wish to produce a reproduction of what the 'Romanized' infantry of the Seleucid army might have looked like, we therefore have to look elsewhere to establish what type of helmet might have been used. It is possible that the Roman-armed infantry wore locally produced versions of the Montefortino helmet in bronze, but this has to be considered as very unlikely. Alternatively they could have worn one of the late Hellenistic styles of helmet prevalent in the Seleucid Empire. We have neither actual examples nor representations of helmets dateable to the late second century BC other than on coins.

The most famous representation of a helmet on Seleucid coinage comes first on a single issue of Antiochus VI and then regularly on coins of Tryphon which show the famous horned helmet which formed his 'badge'.[54] The helmet has a tall skull reinforced by a corrugated crown, above which rises an ornately decorated spike. The skull is decorated with two circular medallions, one showing an eagle astride a thunderbolt, the other showing a panther carrying a thyrsos. Below these medallions is a visor ending with a circular terminal. The visor is decorated with some floral or vegetal design. Below the visor a brim runs round the bottom of the skull, splayed out over the ears in the shape of a 'Boeotian' helmet. Above the front of the brim is a wreath of oak leaves and the single horn of a wild goat. The cheek-

[54] Houghton, *op. cit.* nos. 242, 254-263.

pieces are decorated with a thunderbolt. The helmet is usually considered to be a personal badge of Tryphon, either his own helmet or a symbol of the army that had elected him.[55] Seyrig sought to identify the emblems on the skull of the helmet with the city of Apamea, the eagle and the thunderbolts being symbols of Apamean Zeus.[56] This interpretation has been rejected by Baldus,[57] and indeed it seems best to interpret it as a symbol which belongs to Olympian Zeus. Given the horn on the helmet, which is surely to be interpreted as a symbol of royal divinity, it seems best to identify the helmet as belonging to the monarch himself, first Antiochus VI and then Tryphon. Consequently it must be assumed that it is an elaborate version of the helmet normally worn by the élite Seleucid Companion cavalry regiment. It is extremely unlikely that helmets of this type were used by Seleucid infantry during the period in question.

A late form of the Boeotian helmet is also shown being worn by monarchs on a number of late Seleucid coins. The most important of these is the coin most probably struck in the mint of Seleuceia on Tigris by Timarchus the Milesian. When Demetrius I took the throne from Antiochus V and had the latter killed, Timarchus, then satrap of Babylonia, broke away from the western section of the Empire, and, probably around 162 had himself proclaimed 'Great King' and started to issue his own coins.[58] There are a number of variant dies of this issue, but they all show Timarchus wearing a late form of the Boeotian helmet. According to Tarn both the title assumed by Timarchus and his coinage are imitations of Eucratides of Bactria,[59] who also issued coins in which he is shown wearing a Boeotian helmet.[60] In these issues however, his helmet is adorned with the horns and ears of a bull, adopted as signs of his kingship, while the coins of Timarchus do not show these features.[61]

A series of coins issued by Alexander I Balas show either Athena, or Alexander himself wearing a crested Boeotian helmet.[62] It has been suggested that Balas may be intending to depict himself as Alexander the Great, who wore a Boeotian helmet of this type along with other cavalrymen of the Macedonian army,

---

[55] Eg. Edwyn Robert Bevan, *The House of Seleucus* (London 1902) II p. 231 'The emblem of King Tryphon was the national helmet of the Macedonians'.

[56] Henri Seyrig, *Notes on Syrian Coins* (= *American Numismatic Society, Numismatic Notes and Monographs* no. 119, New York 1950) pp. 8-9.

[57] H. R. Baldus, 'Der Helm des Tryphon und die seleukidische Chronologie der Jahre 146-138 v. Chr.' *Jahrbuch für Numismatik und Geldgeschichte* 20 (1970) pp. 217-239 pl. 16 esp. p. 222.

[58] P. Gardner, *A Catalogue of Greek Coins in the British Museum, Seleucid kings* (London 1878) p. 50; Haughton, *op. cit.* no. 990.

[59] W.W. Tarn, *The Greeks in Bactria and India*^{2} (Cambridge 1951) p. 218.

[60] Eg. Norman Davis and Colin M. Kraay, *The Hellenistic Kingdoms. Portrait Coins and History* (London 1973) pls. 144, 147, 148.

[61] Cf. A. K. Narain, *The Indo-Greeks* (Oxford 1957) p. 54.

[62] Houghton, *op. cit.* nos. 200-203.

Figure 27. Helmet shown on a bronze coin of Askalon (photograph: Cabinet de Médailles, Bibliothčque Nationale, Paris).[63]

as is clearly demonstrated by the 'Alexander' Sarcophagus of King Abdalonymus of Sidon and the Alexander Mosaic from Pompeii.[64] If this interpretation is correct then Alexander Balas is not wearing a helmet then actually in use by the Seleucid army. Even if the interpretation is incorrect it is much more likely that Boeotian helmets would have been worn by the cavalry, as was recommended by Xenophon (*Eq.* 12. 3), than by the infantry. It would be reasonable to assume that in all these examples both Timarchus and Alexander I Balas are choosing to be shown in the cavalry helmet worn by their élite cavalry guards.

Examples of a different type of helmet type are shown on a number of bronze coins struck during the reign of Antiochus VII at Askalon in Palestine.[65] A typical example of one of these coins is shown in Figure 27. The helmet has a rather shallow skull with a wide brim projecting to the front and a nape-guard curling up at the back. Between the skull and brim is a visor outlined by a projecting line at the top and the bottom, ending in a curling terminal in the centre of the skull. At the top of the skull is a ball supporting a crest flowing front and back. Below the skull hang two cheek-pieces. Almost exactly the same type of late Hellenistic helmet is shown on a gem in the Staatliche Münzsammlung in Munich (Figure 28), which has been dated to the end of the third or the beginning of the second century BC.[66] The only

---

[63] Ernest Babelon, *Catalogue des Monnaies grecques de la Bibliothèque nationale, Les Rois de Syrie, d'Arménie et de Commagène* (Paris 1890) no. 1449 pl. xxv, 16.

[64] Cf. Nick Sekunda and Angus McBride, *The Army of Alexander the Great* (London 1984) p. 16.

[65] Houghton, *op. cit.* nos. 818-9.

[66] *Antike Gemmen in Deutschen Sammlungen. I Staatliche Münzsammlung München* (ed. Elfreide Brandt) *Teil 1* (Munich 1970) no. 384 = Petros Dintsis, *Hellenistische Helme* (1986) pl. 8, 3.

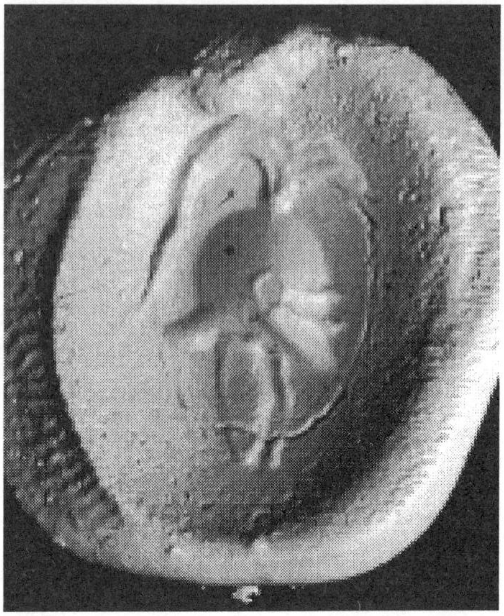

Figure 28. Sealing of helmet on a gem in Munich (photograph: author from sealing).

difference seems to be that the skull of the helmet is deeper. A helmet which is also close in type is shown on a second century agate gem of Campanian-Roman style in the Hague.[67] Here, however, the visor is paralleled by a band running round the back of the helmet between the skull and the brim. Also the cheek-pieces are different, projecting more at the front to cover the chin and cheeks. An actual helmet of this general type has been found in Rumania,[68] demonstrating that helmets of this type are by no means confined to the Seleucid Empire. Perhaps, then, we might postulate that the Roman-armed infantry of the Seleucid phalanx may have worn helmets of this type. Of course there is no guarantee that these helmets are intended for the use of infantry. Nor can we be certain that helmets of this type appear on coins of Askalon because they were worn by the contemporary Seleucid army, rather than for some other reason. Nevertheless this is the only evidence available on the subject.

[67] Marianne Maaskant-Kleibrink, *Catalogue of the Engraved Gems in the Royal Coin Cabinet The Hague. The Greek and Roman Collections* (The Hague 1978) 146 = Dintsis, *op. cit.* pl. 8, 2.

[68] *Die Daker. Katalog der Ausstellungs im Römisch-Germanisch Museum in Köln* (Cologne 1980) fig. 15 = Dintsis, *op. cit.* pl. 56, 6.

Section 2 h

Conclusion

When discussing the Daphne Parade modern scholars have, in the opinion of the author, failed to give due significance to the presence of the 5,000 Roman-armed troops. Most modern authorities pass over their presence without comment, and none see them as the start of a process of Roman reform in the Seleucid army. Thus Launey, though noting that the troops at the Daphne Parade are the first evidence for the adoption of Roman equipment by any Hellenistic army, remarks that they were the work of Antiochus 'ce souverain bizarre et fantasque'. He adopts the view that the Roman-armed infantry of Antiochus are an eccentricity, and that apart from this Hellenistic armies remained faithful to their Hellenic and Oriental traditions right down to the end of the Hellenistic age.[69] Launey sees these troops as an aberration leading to nothing. Similarly Reuter and Mørkholm, though numbered among those modern authorities most favourable to Antiochus, see the 5,000 troops as a 'one-off' creation of Antiochus, not as the start of a process of thorough reform of the infantry.[70] Neither noted the presence of Roman-armed infantry at the battle of Beth-Zacharia. Even Bar-Kochva, who did, sees them as being a one-off experiment of Antiochus rather than being part of a process of on-going reform.[71]

Kuhrt and Sherwin-White are led to comment that 'Polybius was a careful commentator on the armies of the period; it is highly significant that he does not comment on innovations when describing the composition of Seleucid armies, as he could have done had there been any'.[72] This surely is the key. Polybius mentions the presence of the 5,000 Roman-armed infantry in the Parade but fails to pass comment. It is surely the attitude of Polybius to the Seleucid royal house, and to Epiphanes in particular (discussed in Appendix E), which has led him to fail to mention the significance of the Roman-armed infantry in the Parade. This has led modern authorities to likewise pass over the passage without further comment.

It would be easy for the modern scholar, following Polybius, to dismiss Antiochus' innovations as eccentric theatricalities, but in the opinion of the present au-

---

[69] Launey 1 p. 609.

[70] Mørkholm, *Antiochus* p. 39 'Furthermore he equipped part of his army as Roman soldiers'; Reuter, *Beiträge* p. 39 n. 26 'Elitekorps'.

[71] Bar-Kochva, *The Seleucid Army* p. 55; *Judas Maccabeus* p. 315.

[72] Sherwin-White & Kuhrt, *Samarkand to Sardis. A New Approach to the Seleucid Empire* (London 1993) p. 214.

thor this would be a grave mistake.[73] In this chapter it has been suggested that the 5,000 Roman-armed infantry at the Daphne Parade were only the first step in the introduction of Roman organization, equipment and tactics to the whole of the heavy infantry, which took place during the rest of Antiochus' reign. This cannot be proved with certainty, but perhaps the most significant consideration supporting this view is the evidence that the Ptolemaic heavy infantry was thoroughly re-organized in exactly this way during the same decade. The most straightforward interpretation of the description of the Seleucid heavy infantry present at the battle of Beth-Zacharia in 162 given in *2 Macc.* 13. 2 indicates that by that date the rest of the infantry had been reformed along Roman lines.

It is possible that the reform was suggested to the king by one of his officers, but it seems more probable that the reform was introduced upon the initiative of Antiochus himself. Like Polybius, he would have had plenty of opportunity to see the Roman army himself during his period of exile in Rome.

---

[73] Walbank, *Commentary* III p. 286.

GENERAL CONCLUSIONS

The proposition advanced in the second chapter, that the Seleucid heavy infantry was armed and organized along Roman lines at some point immediately prior to 166, is based on two passages in the literary sources and on one disputable piece of archaeological evidence. The material covering the Ptolemaic infantry is much more copious and the conclusion that it was comprehensively reorganized along Roman lines at roughly the same period seems inescapable. It is the certainty of the Ptolemaic evidence which puts the Seleucid evidence in its context.

Thus the conclusion of this work is that both the Seleucid and Ptolemaic heavy infantry were reorganized and equipped along Roman lines during the 160's. Given our lack of a firm *terminus post quem* for the introduction of the maniple into the Ptolemaic infantry it is impossible to know in which army the reform was introduced first. I suspect that Antiochus IV was the first to reform his army, but this contention cannot be supported with evidence.

When Hannibal invaded Italy with less than 20,000 men, Polybius (2. 24. 16) tells us that the Romans and their Allies were capable of mustering, at least on paper, 700,000 foot and 70,000 horse. It was Rome's capacity to mobilize such huge armies which defeated Macedon, rather than any innate superiority of the Roman military system. However many armies the incompetence of Roman military commanders could lose, there was always a near-inexhaustible reservoir of manpower to draw on.[1] The first years of the Third Macedonian War saw many Roman reverses, but these did not matter. All that mattered was the last battle. Looked at from this perspective the efforts of the Seleucid and Ptolemaic states to reform their infantry along Roman lines were doomed to failure. This, however, assumes that both intended to take on the armed might of the Roman state.

It is the view of the present author that modern scholars, too far distanced

[1] Cf. P. A. Brunt, *Italian Manpower 225 B.C. - A.D. 14* (Oxford 1971) passim.

from these events to understand what was going through the minds of the Greeks and Macedonians in the years after Pydna, have dismissed subsequent Hellenistic history as hardly worthy of study, assuming that the ultimate conquest of the eastern Mediterranean by Rome was inevitable. But was it inevitable? More importantly, did the surviving Hellenistic monarchs think it inevitable? The evidence implies that they did not. This is not to suggest that either Antiochus IV or Ptolemy VI thought they could win a war with Rome, but who could foresee what might happen in the future? Over the following year both monarchs made substantial efforts to increase the power of their kingdoms for the future, presumably in the belief that circumstances might change in an entirely unpredictable way. It is generally understood that Hellenistic armies started to use Roman equipment and tactics during the first century BC, but the evidence for military reforms along Roman lines in the early second century has never been properly studied, or even gathered together, previously.

It is a commonplace of military history that successful armies are always imitated, but the imitation is usually quite superficial. Following the Prussian victories of the mid-nineteenth century armies across the world threw away the képi and donned the pickelhaube, but none embarked on the process of thoroughly militarizing their social structures, down to railway timetables, which was the real key to Prussian success. The extent of 'Romanization' of the organizational and tactical structures of the Seleucid and Ptolemaic heavy infantry which took place around the time of the Third Macedonian War, or shortly thereafter, is quite remarkable. Its result was that by the end of the Hellenistic period the heavy infantry of the surviving Hellenistic armies was almost universally equipped and organized along Roman lines. The evidence for this is briefly discussed in Appendix G. Perhaps one might best compare the process of the 'Romanization' of the heavy infantry of the Seleucid and Ptolemaic armies to the westernization of the Ottoman and Egyptian armies during the nineteenth century. The aim of this book has been to present this material to the reader.

APPENDIX A

The Evidence for Roman Influence on Hellenistic Armies before the Third Macedonian War

The first evidence we have for Roman influence of any sort upon the army of a Hellenistic state concerns the Italian Campaigns of Pyrrhus. Pyrrhus lived in a period of rapid political change, the age of the *Diadochoi* 'Successors', after the Empire created by Alexander had broken up. New empires rose and fell with astonishing rapidity as the marshals of Alexander fought for his inheritance. As a result the prime role of the monarch became the provision of leadership in war.[1] At the age of 17, having been expelled from his kingdom, Pyrrhus took refuge in the entourage of Demetrius and fought at the battle of Ipsos. The rest of his life was spent embroiled in one conflict or another.

Plutarch (*Vit. Pyrr.* 8. 3) notes that Pyrrhus was fascinated with warfare and little else:

> καὶ ὅλως τοῦτο μελετῶν ἔοικε καὶ φιλοσοφῶν ἀεὶ διατελεῖν ὁ Πύρρος, ὡς μαθημάτων βασιλικώτατον, τὰς δὲ ἄλλας γλαφυρίας ἐν οὐδενὶ λόγῳ τίθεσθαι. λέγεται γὰρ ὡς ἐρωτηθεὶς ἔν τινι πότῳ, πότερον αὐτῷ φαίνεται Πύθων αὐλητὴς ἀμείνων ἢ Καφισίας, εἰπεῖν ὅτι Πολυσπέρχων στρατηγός, ὡς ταῦτα τῷ βασιλεῖ ζητεῖν υόνα καὶ γινώσκειν προσῆκον.

> Pyrrhus would seem to have been always and continually studying and meditating upon this one subject, regarding it as the most kingly branch of learning, the rest he regarded as mere accomplishments and held them in no esteem. We are told for instance, that when he was asked at a drinking-party whether he thought Python or Caphisias the better flute-player, he replied that Polysperchon was a good general, implying that it became a king to investigate and understand such matters only.[2]

Consequently Pyrrhus was an accomplished general. Plutarch (*Vit. Pyrr.* 8. 2) records that Antigonos, when asked who was the best general among the generation who

---

[1] M. M. Austin, 'Hellenistic Kings, War, and the Economy' *CQ* 36 (1986) pp. 450-66.

[2] Text and translation after the Loeb edition of Bernadotte Perrin.

survived Alexander replied 'Pyrrhus, if he lives to be old'. Hannibal rated Pyrrhus as the best general that had lived. Indeed Pyrrhus was a military intellectual: he is, after all, known to be the author of a number of works on warfare,[3] and was also responsible for a number of military developments, for example the entrenched camp, which was adopted by the Romans after the battle of Beneventum (Front., *Strat.* 4. 1. 14). One of Pyrrhus' most important innovations seems to have been in the realm of battlefield tactics.

Ever since the Macedonian phalanx first emerged as a tactical formation, there existed a danger that the frontage would rupture if the phalanx was required to move forward for any distance over rough terrain. The first evidence for this occurring was at the battle of Gaugamela. The left wing of the Macedonian army had been pinned down by a Persian attack, but Alexander was simultaneously moving forward with the right wing. The pike battalions stationed on the right followed Alexander forward, but the *taxis* commanded by Simmias was no longer able to keep up and maintain the line (Arr., *Anab.* 3. 14. 4). It seems that Pyrrhus was the first commander to attempt to solve this problem by alternating blocks of pike-men and medium infantry in the front line. According to Polybius (18. 28. 10) Pyrrhus made use of Roman weapons and of Italian troops, placing a maniple (σημαία) of Italians alongside a phalanx block (σπεῖρα) in his battles with the Romans.[4] It would be possible to interpret this passage in Polybius as an indication that Pyrrhus had wholeheartedly adopted Roman manipular tactics. Polybius, however, does not say that Pyrrhus introduced manipular tactics, leaving spaces between the maniples. Rather he says that he interspaced maniples of Italians between blocks of pikemen.[5] The maniples of Italians would have served as flexible 'joints' between the pike-blocks, enabling the latter to deliver a series of individual 'hammer-blows' without disrupting the line. In this way the phalanx could 'articulate' and a rupture in the line was prevented. We might call this formation an 'articulating phalanx'. It would be incorrect to say that Pyrrhus changed the tactics of his army to those used by Rome. Rather his tactical innovations were probably inspired by Italian manipular tactics.[6]

---

[3] Plut., *Vit. Pyrr.* 8. 2; Cic., *Epist. ad div.* 9. 25; Müller, *FHG* ii p. 461; Cf. Everett L. Wheeler, *Stratagem and the Vocabulary of Military Trickery* (*Mnemosyne Suppl. 108*, Leiden 1988) p. 14 n. 40.

[4] Cf. Pierre Lévèque, *Pyrrhos* (= *Bibliothéque des Écoles françaises d'Athénes et de Rome Fasc. 185*, Paris 1957) p. 542.

[5] Following Walbank, *Commentary* II p. 586.

[6] Eg. Johannes Kromayer, Georg Veith, *Heerwesen und Kriegführung der Griechen und Römer* (= edd. Iwan von Müller, Walter Otto, *Handbuch der Altertumswissenschaft IV, 3, 2*, Munich 1928) p. 263. Some modern authors maintain that the exchange was the other way round, in other words that Pyrrhus was the inventor of manipular tactics, which the Romans subsequently copied (Cf. Paul Couissin, *Les Institutions militaires et navales* (Paris 1932) p. 83: 'On a pensé que la tactique de Pyrrhus avait inspiré le formation manipulaire de la légion romaine', and on p. 120: 'Enfin, c'est peut-être d'un Grec, le roi Pyrrhus, que Rome avait appris la tactique qui vainquit la Grèce').

As has been noted by Walbank, the same type of 'articulating phalanx' was employed by Antigonus Doson at Sellasia in 222 BC, where he interspersed sub-units (σπεῖραι) of his *Chalkaspides* and Illyrians (Polyb. 2. 66. 5). The *Chalkaspides*, or 'Bronze-shields' were one of the regiments of the Antigonid phalanx. The Macedonian line had to march up the steep slope of the Euas and once again the purpose of this 'articulating' phalanx line was that the lighter-armed and more flexible Illyrians would ensure that no gaps opened up in the phalanx. Walbank noted relative elasticity of this formation 'which had something of the flexibility of the Roman manipular organization, contrasted with the normal phalanx. Antigonus was following the lesson of Pyrrhus, who adapted his army to Italian arms and methods'.[7]

According to Polybius (11. 11. 6) at the battle of Mantineia in 207,[8] fought against the Lakedaimonian army commanded by Machanidas, Philopoemen stationed the Achaean phalanx 'by divisions in *speirai* at intervals' (κατὰ τέλη σπειρηδὸν ἐν διαστήμασιν ἐπέστησε). During the battle these divisions were able to wheel about to the left (11. 15. 2). Polybius sometimes uses the word σπεῖρα to translate the Roman terms maniple (3. 115. 6) and sometimes cohort (11. 23. 1), but it is also the standard word used to describe a phalanx block of 256 men in the Macedonian army.[9] Philopoemen had just earlier re-armed the phalanx of the Achaean army as phalangites in the Macedonian manner with pikes (*sarisai*), shields, helmets and cuirasses.[10] It seems best to conclude that the basic unit of the Achaean phalanx was now termed a *speira* and that Polybius is using the word σπειρηδὸν in the sense 'in pike blocks' rather than implying that Philopoemen had adopted manipular tactics wholesale.[11] The fact that gaps were left between the σπεῖραι is, however, significant, and it is possible that Philopoemen was copying Roman tactics.[12] Nevertheless the weapons used by both the Achaean and the Macedonian phalanx remained the traditional ones.

An 'articulating' phalanx also seems to have been the tactical formation used

---

[7] Walbank, *Commentary* I p. 280.

[8] For general treatments of this battle see Walbank, *Commentary* II p. 283. Topographical and other problems concerning the battle have been dealt with in W. Kendrick Pritchett, *Studies in Ancient Greek Topography Part II (Battlefields)* (Berkeley and Los Angeles 1969) 37-72.

[9] F. W. Walbank, *Philip V of Macedon* (Cambridge 1940) p. 293. Feyel (*Rev. arch.* 1935, pp. 45-7 demonstrates that although the term σπεῖρα eventually comes to be used of the Roman cohort and even of religious associations in the Imperial period, the word is originally used of a Macedonian pike-block of 256 men. For a wide-ranging discussion of the various meanings of this term see the article of Poland, 'Σπεῖρα' *PW* iii A, 2 (1929) col. 1586-1592.

[10] Plut., *Vit. Phil.* 9. 1-7; cf. Paus. 8. 50. 1; R. M. Errington, *Philopoemen* (Oxford 1969) pp. 64-5.

[11] Cf. Walbank, *Commentary* II p. 286.

[12] Paul Couissin, *Les Institutions militaires et navales* (Paris 1932) p. 83 speaks of Philopoemen arranging his phalanx in battalions in chequer formation, so Coussin clearly thinks he was copying Roman tactics.

at the battle of Magnesia in 190 BC, where Antiochus the Great deployed the pikemen of the Seleucid army as follows:

> Decem et sex milia peditum more Macedonum armati fuere, qui phalangitae appellabantur. Haec media acies fuit, in fronte in decem partes divisa; partes eas interpositis binis elephantis distinguebat; a fronte introrsus in duos et triginta ordines armatorum acies patebat.
>
> There were 16,000 infantry armed in the Macedonian manner, who are called phalangites. They formed the centre of the line, and their frontage was divided into 10 parts; and these parts were separated by intervals in which two elephants each were placed; from the front the formation extended 32 ranks in depth (Livy 37. 40. 1).[13]
>
> ... ἡ φάλανξ ἡ Μακεδόνων ἄνδρες ἑξακισχίλιοι καὶ μύριοι, ἐς τὸν 'Αλεξάνδρου καὶ Φιλίππου τρόπον ἔτι κοσμούμενοι· ἵστη δ'αὐτοὺς ἐν μέσῳ, διελὼν ἀνὰ χιλίους καὶ ἑξακοσίους ἐς δέκα μέρη, καὶ τούτων ἑκάστου μέρους ἦσαν ἐπὶ μὲν τοῦ μετώπου πεντήκοντα ἄνδρες, ἐς δὲ τὸ βάθος δύο καὶ τριάκοντα, ἐς δὲ τὰ πλευρὰ ἑκάστου μέρους ἐλέφαντες δύο καὶ εἴκοσιν. ἡ δ' ὄψις ἦν τῆς μέν φάλαγγος οἷα τείχους, τῶν δ' ἐλεφάντων οἷον πύργων. τοιοῦτων μὲν ἦν τὸ πεζὸν 'Αντιόχου.
>
> ... the Macedonian phalanx of 16,000 men, still arrayed after the fashion of Alexander and Philip. These were placed in the centre, divided into ten *mere* of 1600 men each, with 50 men in the front line of each section and 32 deep. On the flanks of each section were 22 elephants. The appearance of the phalanx was like that of a wall, of which the elephants were the towers. Such was the arrangement of the infantry of Antiochus (Appian, *Syr.* 32).[14]

Appian misinterpreted the sources he made use of, which in this case seems to have been Polybius.[15] Presumably there were 2 elephants in each gap, which would make 22 in all if both flanks of the phalanx were covered by elephants.[16] Each of

---

[13] Text and translation after the Loeb edition of Evan T. Sage.

[14] Text and translation after the Loeb edition of Horace White.

[15] On this misunderstanding of Appian see Kai Broderson, *Appians Antioche (Syriake 1, 1 - 44, 232) Text und Kommentar nebst ein Anhang: Plethons Syriake - Exzerpt etc.* (Munich 1991) p. 178 'ist Appians Darstellung Sy 162 mißverständlich formuliert; zweifellos waren nicht jeweils, sondern insgesamt 22 Tiere, je 2 an den beiden Außenseiten und in den 9 Zwischenräumen aufgestellt'.

[16] Bar-Kochva, *Seleucid Army* p. 263 n. 18;

the elephants would have been accompanied by a guard of an uncertain number of infantry.[17] Once again the purpose of this 'articulated' formation of the phalanx seems to have been to impart some degree of flexibility to the Seleucid phalanx. In this case ever more necessary in view of the fact that it was about to join battle with the highly mobile legionary formation.[18]

We have seen that it was reasonably common practice for the phalanxes of third century Hellenistic armies to be drawn up in articulating phalanx blocks until the battle of Magnesia. It seems that this 'articulating phalanx' was a military innovation of Pyrrhus, the result of his experiences in the Italian Campaign. Pyrrhus was clearly influenced by manipular tactics. There is no indication, however, that later armies used weapons other than the traditional ones of the phalanx. Also there is no reason to believe that the phalanx blocks themselves were deployed in the open formations adopted by Roman heavy infantry. Rather they were drawn up in close order, normally interspersed with medium infantry drawn up in looser formation which allowed the phalanx to articulate. In these cases it seems probable that these formations were adopted as a result of Pyrrhus' military writings and we should not regard them as a tactical reform imitating contemporary Roman military practice. An exception to this may be Philopoemen's formation at Mantineia, where gaps were left between the phalanx blocks. There is no indication that the phalanx blocks themselves were drawn up in loose order however, and we do know that the weapons they used were the Macedonian *sarisa* and heavy armour.

It therefore seems possible that there may be some connection between manipular tactics and the evolution of the 'articulating phalanx' of Hellenistic armies. The concept was probably first developed by Pyrrhus under the influence of manipular tactics. Rather than an adoption of manipular tactics by a Hellenistic army it constituted an adaptation of manipular tactics to the Hellenistic phalanx. The concept of introducing gaps between pike blocks and interspersing sub-units of lighter infantry between pike blocks was new, but the close order and the equipment used by the pikemen remained traditional. Consequently the evidence indicates that any Roman influence on Hellenistic infantry tactics and organization before the Third Macedonian War was superficial and would not justify the hypothesis that any Hellenistic army attempted to reform the organization and tactics of its infantry along Roman lines.

We might now return to the statement of Polybius (18. 28. 10) that Pyrrhus made use of Italian weapons as well as Italian forces. It is not clear whether Polybius means that the Italian weapons Pyrrhus used were in the hands of Italians, or whether he means that Pyrrhus gave Italian weapons to his own Epirote troops. It

---

[17] Bar-Kochva, *Seleucid Army* pp. 8, 82, 167.

[18] Cf. J. Kromayer, G. Veith, *Antike Schlachtfelder in Griechenland* 2 (Munich 1907) pp. 214-6 who comments on the intervals in the phalanx both at Magnesia and at Mantineia, quoted in Bar-Kochva, *Seleucid Army* p. 263 n. 17.

Figure 29. Cheek-pieces from a Montefortino helmet found at Dodona, National Museum Athens, Carapanos Collection 158 and 170 (photograph: National Museum, Athens).

is worth noting some finds of Italian military equipment on the Greek mainland which could be connected with the Epirote army.

A pair of cheek-pieces has been recovered at Dodona (Figure 29), the national shrine of Epirus, during excavations carried out in the last century by Carapanos.[19] They are probably from a Montefortino Type A helmet following Robinson's typology, and so date to the fourth or third centuries BC.[20] We do not know if they represent a dedication of a helmet by its wearer or a dedication of booty. It would be going beyond the evidence to suggest that they once belonged to an Epirote soldier, and that some of the Epirote army wore Montefortino helmets. Another early example of a helmet of Montefortino type was found at Zagora Polydrossou in 1910 (Figure 30). The publisher dated the helmet to the third quarter of the third century BC and suggested that 'most probably it has been brought to Greece by a soldier of Pyrrhus' army returning from the Italian expedition'.[21]

A famous bust (Figure 31) has been identified as showing Pyrrhus himself. It depicts a warrior wearing a helmet of indeterminate type. The front part of the helmet is restored.[22] The identification with Pyrrhus is based on the simple fact that the style points to a date in this period, and that the oak wreath that the bust wears

---

[19] Constantin Carapanos, *Dodone et ses Ruines* (Paris 1878) pl. LVI, 6.

[20] H. Russell Robinson, *The Armour of Imperial Rome* (New York 1975) p. 17.

[21] Alk. Spetsieris, 'Early Roman Bronze Helmet from Polydrosso in Thresprotia' *AAA* 15, 1 (1982) pp. 59-64.

[22] G. M. A. Richter, *The Portraits of the Greeks. Volume Three* (London 1965) p. 258; G. M. A. Richter, abridged and revised by R. R. R. Smith, *The Portraits of the Greeks* (Oxford 1984) p. 229-30.

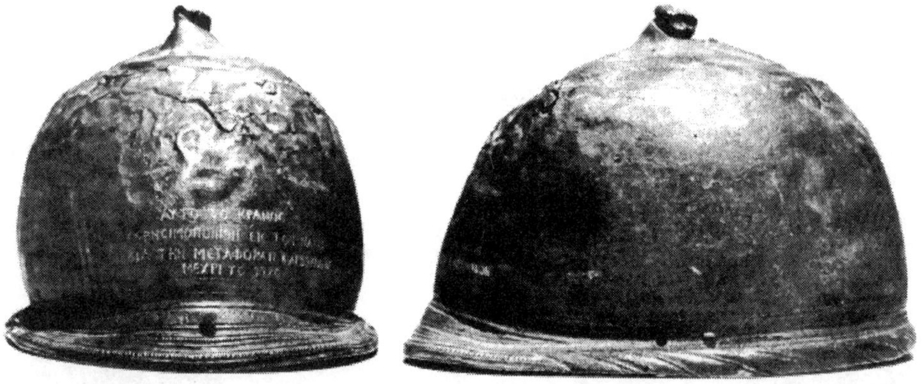

Figure 30. Montefortino helmet found at Zagora Polydrossou in 1910 (photographs after *AAA* 15, 1 (1982) pp. 59-64).

is the national symbol of Epirus, derived from the oak at Dodona, though this identification is itself uncertain. This bust was used by Russell Robinson to establish the precise way in which the cheek-pieces of a Montefortino helmet were kept together. 'Secured over external studs on the base of each cheek-piece are narrow straps with stepped terminals which are crossed over at the chin, passing closely under the jaw at each side and round to the back of the warrior's neck, where they pass beneath the neck-guard'.[23] By comparison with Montefortino helmets it can be assumed that these straps were secured by two bronze rings fitted to the back of the helmet. The helmet is not of Italian type, nor are the cheek-pieces of the normal Montefortino type. The Celtic or Italian system of fixing the cheek-pieces on this helmet could be significant, and could support Polybius' statement that Pyrrhus made use of Italian weapons, if we could be sure that the bust shows Pyrrhus himself.[24]

These various pieces of archaeological evidence lend some weight to Polybius' assertion that Pyrrhus made use of Italian weapons as well as tactics. Nevertheless they do not indicate that the Epirote army adopted Italian equipment wholesale. There is no evidence that any other Hellenistic state of the eastern Mediterranean adopted Roman equipment for its infantry.

There is some evidence that the Carthaginian army adopted Roman military practices, in this case as a direct result of contact with the Roman army. Polybius

---

[23] H. Russell Robinson, *The Armour of Imperial Rome* (New York 1975) p. 14.

[24] Cf. Rolf Winkes, 'The Pyrrhus Portrait' in edd. Tony Hackens, Nancy D. Holloway, R. Ross Holloway and Ghislane Moucharte, *The Age of Pyrrhus* (*Archaeologia Transatlantica* XI, Louvain-La-Neuve, 1992) pp. 175-88 who suggests that this bust, if it indeed does show Pyrrhus, shows him wearing it as spoil of war from his victory over the Celts.

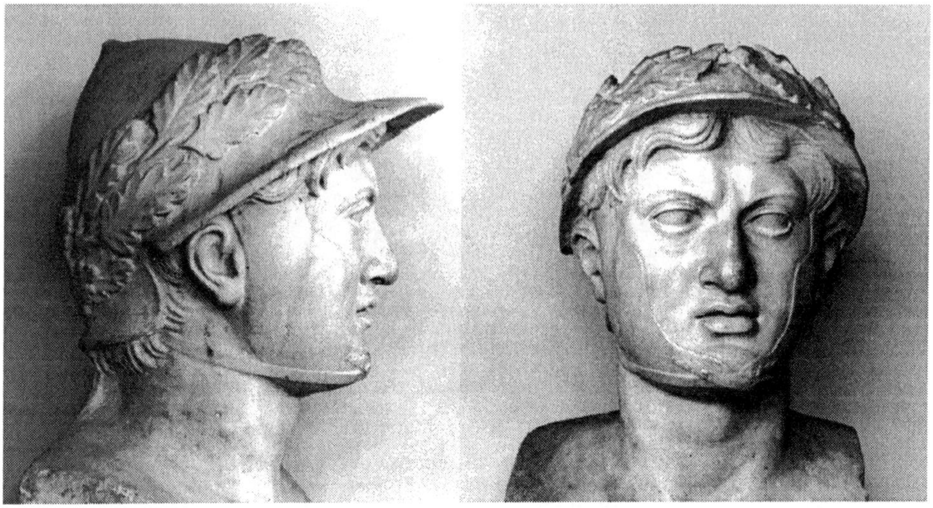

Figure 31. Bust identified as Pyrrhus, National Museum, Naples inv. 6150 (photograph: German Archaeological Institute, Rome).

(18. 28. 9) notes that Hannibal armed his own forces with Roman weapons after his first victory and continued to use them right up to the end. This occurred after the battle of Lake Trasimene, when Hannibal found himself in possession of a large quantity of captured Roman equipment (Polyb. 3. 87. 3). The next year the Libyan contingent at Cannae was armed in the Roman fashion 'Hannibal having equipped them with the choicest of arms captured in the previous battles' (Polyb. 3. 114. 1). Finally at Zama Hannibal's rear line of veterans were seemingly still equipped in the Roman manner, as Polybius (15. 14. 6) tells us that both sides were equally well armed.[25] There is no suggestion, however, that the Libyan infantry fighting for Hannibal adopted Roman tactics in addition to their equipment.

The evidence, both literary and archaeological, indicates that even though we can detect Roman influence on Hellenistic infantry organization and equipment before the Third Macedonian War, the effects of this influence were sporadic and did not amount to changes which could be described as a complete reform of the infantry along Roman lines.

[25] Cf. Walbank, *Commentary* II p. 586.

APPENDIX B

The Taktika of Poseidonius of Apameia

A variety of military literature has survived from classical Antiquity, one branch of which is known as the *Taktika* which is represented by three works each by different authors called Asclepiodotus, Aelian and Arrian, describing the organization of a Greek army. The contents of all these three works are very similar, and the relationship of one to another is disputed.

The oldest is the Τέχνη τακτική of Asclepiodotus. As will be seen below, although the identity and therefore the date of Asclepiodotus cannot be established with complete certainty, he seems to date to the early first century BC. It is generally agreed that Asclepiodotus' work was closely based on the earlier Τέχνη τακτική of the famous stoic philosopher, Poseidonius of Apamea (otherwise known as Poseidonius Rhodius).[1] The precise relationship between these two figures will be investigated below.

The Τακτική θεωρία of Aelian was written considerably later. The identity of Aelian is completely unknown,[2] but a date of about AD 110 can be established for his *Taktika* by its dedication to the Emperor Trajan.[3] Given this date, it is certain that Aelian cannot be identified with Aelianus Claudius, the writer of the *De natura animalium* and the *Varia Historia*, whose dates are ca. AD 170 to ca. AD 235.[4] The treatise of Aelian is close to that of Asclepiodotus in its organization and contents, but it goes into considerably more detail, and has whole sections, such as the introduction listing previous writers of *Taktika*, which are completely absent in Asclepiodotus. It is thus possible that Aelian based his treatise on that of Asclepiodotus,[5] supplying material absent in the latter from his own researches. It seems more likely, on the other hand, that Aelian has gone back to the original, fuller, treatise of Poseidonius, which Asclepiodotus epitomized, and has made his own

---

[1] Tarn, *OCD*^{2} sv. 'Asclepiodotus' p. 129.

[2] Cf. K. K. Müller, 'Aelianus (10)' *PW* i. 1 (1894) 482-6.

[3] This date is given by Philip A. Stadter, *Arrian of Nicomedia* (Chapel Hill 1980) p. 41, though Everett Wheeler, 'The Occasion of Arrian's *Tactica*' *GRBS* 19 (1978) pp. 351-365, esp. p. 353 gives Aelian's work a date of AD 101. See also p. 351 n. 1 for a bibliography of previous works on the *Tactica.*

[4] Cf. *Aelian, Historical Miscellany, Edited and Translated by N. G. Wilson* (Loeb ed. Cambridge Mass., London 1997) p. 2.

[5] Thus James G. DeVoto (ed., tr.). *Flavius Arrianus ΤΈΞΝΗ ΤΑΚΤΙΚΆ(Technical Handbook) and ἘΚΤΑΞΙΣ ΚΑΤᾺ 'ΑΛΆΝΩΝ(The Expedition Against the Alans)* (Chicago 1993) p. ii.

abridgement.[6] It is significant that the latest entry in the list of writers in Aelian is Poseidonius, indicating that Aelian based his work on that of Poseidonius, whereas Asclepiodotus is not mentioned in the list. If this supposition is correct, how faithful and complete an abridgement of Poseidonius the *Taktika* of Aelian represents is anybody's guess. It is unfortunate that the identity of Aelian cannot be established, for his treatise seemingly remained influential for a quarter of a century or more until it was used by Arrian as a basis for his own work.

The Τέχνη τακτική of Arrian was composed in AD 136/7 while Arrian was serving as governor of Cappadocia.[7] His treatise consists of two parts. The first is closely modelled on that of Aelian. It sometimes repeats passages from Aelian wholesale, and where Aelian differs from Asclepiodotus, for example in describing the organization of the phalanx at its highest level, Arrian follows Aelian rather than Asclepiodotus. Consequently it is hard to avoid the conclusion that Arrian used Aelian directly, rather than Aelian's source, in this section of his work, although some scholars hold the opinion that both drew on a common source.[8] The second part of Arrian's *Technê Taktikê* constitutes a completely original section describing contemporary Roman cavalry exercises.[9]

Thus we have Asclepiodotus who, though giving a briefer epitome, is closer in time to Poseidonius, and Aelian, who, though more distant in time, probably went back to Poseidonius directly. It cannot be established whether material in Aelian but not found in Asclepiodotus goes back to Poseidonius or is material added to the

---

[6] Cf. H. Tonnet, *Recherches sur Arrien. Sa Personalite et ses ecrits atticistes* (Amsterdam 1988) I p. 92 'Posidonius, la source ultime de la *Tactique* d'Elien'.

[7] The *Ars Tactica* is dated to AD 136/7 by reference to Hadrian's twentieth regnal year at 44. 3 and Wheeler, *op. cit.* 354 suggests that the treatise was written for Hadrian's *vicennalia*, the twentieth anniversary of his assumption of tribunician power in AD 136. On Arrian in general see the works cited in A. B. Bosworth, *From Arrian to Alexander. Studies in Historical Interpretation* (Oxford 1988) p. 16 n. 1.

[8] Thus Philip A. Stadter, *Arrian of Nicomedia* (Chapel Hill 1980) p. 41 'All three seem to derive independently from a common source, for although there are various differences they frequently repeat each other verbatim. Arrian and Aelian are especially close in certain sections', and Philip A. Stadter, 'The *Ars Tactica* of Arrian: Tradition and Originality' *CP* 73 (1978) pp. 117-128 esp. 121 'The modern reader is surprised to discover that even the introductory section, which reviews earlier writers of tactical manuals, contains the same names as Aelian, in the same order'. Cf. A. B. Bosworth, 'Arrian's Literary Development' *CQ* n.s. 22 (1972) pp. 163-185 esp. p. 183. Brian Campbell, 'Teach Yourself How to be a General' *JRS* 77 (1987) pp. 13-29, esp. p. 18 is undecided as to whether Arrian used Aelian or whether both used a common source. Wheeler *GRBS* 19 (1978) p. 353 n. 9 maintains that Arrian drew directly on Aelian, an opinion shared by DeVoto, *op. cit.* p. i. Cf. H. Tonnet, *Recherches sur Arrien. Sa Personalite et ses ecrits atticistes* (Amsterdam 1988) I p. 90 'La *Tactique* d'Elien, qu'Arrien démarque'.

[9] This section has been translated into German by Franz Kiechle, "Die 'Taktik' des Flavius Arrianus", *Bericht der römisch-germanischen Kommission* 45 (1965) pp. 87-129, and into English, along with assorted other sections of the work, in Ann Hyland, *Training the Roman Cavalry. From Arrian's Ars Tactica* (1993) pp. 69-77.

epitome by Aelian himself. For this reason Asclepiodotus has been taken as a model for what might have been in Poseidonius originally in this present work. It has sometimes been necessary to refer to Aelian and Arrian where they differ from Asclepiodotus and it is more likely that they contain a better version of what was originally to be found in Poseidonius.

Poseidonius, generally known as 'the Rhodian', was born in northern Syria, at Apamea on the Orontes about 135 BC.[10] Poseidonius studied at Athens under the Stoic philosopher Panaitios at some point after Panaitios' elevation to the position of head of the 'Stoa' in 129, but before Panaitios' death in 109.[11] Poseidonius then travelled widely carrying out scientific research before settling in Rhodes, which became his adoptive home. Poseidonius was not simply a contemplative philosopher: he also wrote many works of a more practical nature, and he is perhaps best known for his *Histories*, written in 52 books, which continued the history of the growth of Rome from the point where Polybius broke off down till the dictatorship of Sulla in 81. Another of his practical works was his Τέχνη τακτική, 'The Art of Tactics', which is now lost. He died around 51. In the history of ancient thought Poseidonius yields pride of place only to Aristotle.

Little is known of the early life of Poseidonius, and we are ignorant of the precise date of his arrival in Athens.[12] Poseidonius would have entered his eighteenth year and become liable for ephebic training, and then for military service in the Seleucid army, around the year 117.[13] He would have had ample opportunity to see military service. Chronic civil war had been raging in Syria between the two branches of the House of Seleucus. Antiochus VIII Grypos, a descendant of Demetrius I, had just seized power from Alexander Zabinas, a descendant of Antiochus IV Epiphanes, whom he had poisoned about 121. The kingdom then saw some peace for a few years, during which period Grypus, who had been a student in Athens himself, spent his time composing verses on the subject of poisonous snakes. In 116, though, came the invasion of Antiochus IX 'The Cyzicene', the next representative of the other branch of the Seleucid house, and civil war broke out in Syria once again. A general named Herakleon eventually took command of the army of

---

[10] The conventional dates of 135-51 for Poseidonios' life are very insecure (cf. I.G. Kidd, *Posidonius II. The Commentary (i)* (1988) 8-9).

[11] Dr. Modestus Van Straaten O.E.S.A., *Panétius, sa vie, ses écrits et sa doctrine avec une édition des fragments* (Amsterdam 1946) pp. 376-7 gathers the testimonia for Poseidonius being a student of Panaitios. He puts the birth of Poseidonius in 135 and the death of Panaitios a little before 109 or 108 (p. 25).

[12] Cf. eg. Marie Laffranque, *Poseidonios d'Apamée. Essai de Mise au Point* (Paris 1964) 56-7; cf. Karl Reinhardt, 'Poseidonios (3)' *PW* xxii, 1 (1953) 563 'um 110'..

[13] Some seven ephebes from Antioch on the Orontes are listed in an inscription from Delos. Rather than regarding them as having opted to go through the ephebate outside Syria, it is better to look upon them as having been drawn from the foreign population resident on the island (P. Roussel, 'La population de Délos a la fin du IIe. siècle av. J. -C.' *BCH* 55 (1931) pp. 438-449, esp. p. 447.

Antiochus Grypus. In 96 BC Herakleon had Grypus murdered and took the throne, carving out a kingdom for himself centred around his native city of Beroia in north-eastern Syria.[14]

We have no knowledge as to whether Poseidonius came from a military family, though it is possible that he was related to the Poseidonius mentioned as one of the three officers sent by Nikanor 'the Elephantarch' to negotiate with Judas Maccabaeus in 161.[15] Poseidonius' father must have been wealthy in order to have paid for his studies in Athens, and so he was presumably prominent in local life. The surviving fragments of Poseidonius make it clear that his political sympathies lay with the descendants of Antiochus IV, though we have no knowledge of whether he took an active part in the politics, or even the military campaigns, of the day.[16] Modern historians, after noting that the native city of Poseidonius was the central military depôt of the Empire, have suggested that Poseidonius must have received his first education in the state gymnasium in Apamea.[17] This would have included physical education and basic military training as well as academic study. Poseidonius was given the nick-name of 'The Athlete'. We do not know whether this is a reference to Poseidonius' formidable mental gymnastics, or whether he had, indeed, followed an athletic career in his youth.[18] It would presumably therefore have been in his home city that Poseidonius acquired his first knowledge of, and interest in, military matters, which later manifested themselves not only in *The Art of Tactics*, but also in the detailed descriptions he gives of the weapons of the various peoples of the ancient world in the preserved fragments of his works.[19] Even in the unlikely event that Poseidonius did not perform his military service, it would be logical to assume that the theoretical army whose organization was described by Poseidonius was based, even if only loosely and in part, upon the Seleucid army in the last decades of the second century BC. In summary, then, there is no reason to doubt that the work of Poseidonius was an entirely original work, based on his own knowledge and experience.

Polybius (9. 20. 4) mentions that he wrote a work called *Notes on Tactics* (Ὑπομνήματα περὶ τὰς τάξεις).[20] Polybius held the post of hipparch of the Achaean League in 170/69, and scholarly opinion is divided as to whether Polybius

---

[14] John D. Grainger, *The Cities of Seleukid Syria* (Oxford 1990) pp. 172-4.

[15] *2 Macc.* 14.19; I. G. Kidd, *Posidonius II. The Commentary (i)* (1988) p. 6 quoting I. Bake, *Posidonii Rhodii Reliquae Doctrinae* (Leiden 1810) p. 255.

[16] Jürgen Malitz, *Die Historien des Poseidonios* (= *Zetemata* 79, Munich 1983) p. 8.

[17] Eg. Malitz, *op. cit.* pp. 8-10.

[18] Kidd, *op. cit.* p. 3.

[19] Malitz, *op. cit.* 9-10.

[20] For a hypothetical attempt to reconstruct the basic themes of this work see Lucien Poznanski, 'Essai de Reconstruction du *Traité de Tactique* de Polybe d'après le Livre III des *Histoires*' *L'Antiquité classique* 49 (1980) pp. 161-172.

composed this work before or after he held this office.[21] All that can be stated for certainty is that it obviously pre-dates the commencement of work on the *Histories*.[22] This work on Tactics was known to both Aelian (*Tact.* 1. 3-4, 3. 4, 19. 10) and Arrian (*Tact.* 1. 1) and this has led to suggestions that the ultimate model for the *Taktika*, including the lost work of Poseidonius, may have been the work of Polybius.[23] When, however, Aelian quotes Polybius on the cavalry formation known as the rhomb, it is after Aelian has given his description of the various forms a rhomb can take and that it should number 36, and it is because Polybius' description of a rhomb as being 64 men in the shape of the Greek letter *lambda*, is different from the one just given. This in fact clearly proves that Polybius was not used as a model: he is quoted by way of contrast. Elsewhere Aelian (3. 4) first of all quotes his definition of what '*Taktika*' are concerned with, and then quotes the definitions given to *Taktika* by both Aeneas and Polybius. Once again, Polybius is quoted in contrast. Polybius himself mentions that he discusses the utility of a knowledge of geometry in changing the plan of a camp in his work, but it undoubtedly would have concerned itself with, among other things, the different formations to be adopted by an army. All works dealing with *Taktika* concerned themselves with these matters: we know of a multitude of such works and all of them would be similar to that extent. The question is, however, who wrote the treatise which formed the basis for the works of Asclepiodotus, Aelian and Arrian. The fact that the introduction contained in both Aelian and Arrian, which gives a list of previous writers of *Taktika* including Aeneas and Polybius, ends with Poseidonius is a firm indication that it was upon this work that Aelian and Arrian based theirs. Pédech was surely correct in rejecting the proposal that Asclepiodotus was based on Polybius.[24]

Asclepiodotus, 'The Philosopher' as he is called in the heading to his treatise, is generally identified with an individual mentioned a number of times in the works of Seneca, who at one point (*Quest. Nat.* 2. 26. 6) is called an '*auditor Posidonii*': that is, one who had listened to Poseidonius' lectures.[25] This Asclepiodotus is not,

---

[21] The date of the composition is discussed by Walbank, *Commentary* II p. 148. H. M. Werner, *De Polybii vita et itineribus quaestiones chronologicae* (Diss. Leipzig 1877) pp. 15-16 (non vidi) suggested that it was written in Rome after he had held office, a suggestion which is accepted by F. Susemihl, *Geschichte der griechischen Literatur in der Alexandrinerzeit* II (Leipzig 1892) p. 90 n. 48 who would put its composition in the very first years of Polybius' period of exile, 'but it is perhaps more probable that he wrote the *Tactics* in the 170's before he became hipparch' (Walbank, *loc. cit.*) as was suggested by F. Ziegler, 'Polybios (1)' *PW* xxi, 2 (1953) col. 1473. Paul Pédech, *Polybe, Histoires I* (ed. Budé 1969) p. xiii thought that Polybius composed the treatise at Rome during his exile and founded his reputation as a military expert.

[22] Cf. F. W. Walbank, *Polybius* (Berkeley, Los Angeles, London 1972) p. 15 'unless the passage in book ix which mentions it is a later insertion - and there is no reason to think this - it preceded the *Histories*'.

[23] K. K. Müller, 'Asklepiodotus (10)' *PW* ii (1896) col. 1640-41; Tarn *OCD*^{2} p. 129.

[24] Paul Pédech, *Polybe, Histoires I* (ed. Budé 1969) p. xiii.

[25] Eg. Oldfather in *Aeneas Tacticus, Asclepiodotus, Onasander etc.* (Loeb ed., Cambridge Mass.

however, called a disciple (*discipulus*) of Poseidonius, so the identification is not certain.[26] We know of another philosopher of roughly this period called Asclepiodotus son of Asclepiodotus of Nicea, who is listed as a pupil of Panaitios in a papyrus from Herculaneum which mentions that he 'also visited Rome'.[27] In the main modern commentators hold that it is impossible for these two to have been the same individual on account of chronological considerations, and that our Asclepiodotus is probably a later homonym.[28] It is possible, though, that Asclepiodotus son of Asclepiodotus of Nicea and Poseidonius were contemporaries, or at least near contemporaries, and that both studied at Athens under Panaitios in the decade before the latter's death in 109. This Asclepiodotus could have been an *auditor* of Poseidonius at some later date, presumably in Rhodes. Perhaps, then, Asclepiodotus' treatise may have been released subsequently, perhaps during his period of residence in Rome. This would suggest that the *Art of Tactics* may well have been one of the earlier works of Poseidonius. The precise relationship between the *Art of Tactics* of Poseidonius and Asclepiodotus is unknown. It could be held that Asclepiodotus simply copied Poseidonius, but there would be no evidence whatsoever for such a view. It seems more likely that we have a paraphrase or epitome of Poseidonius' earlier work.

Whoever 'Asclepiodotus the Philosopher' was, and whatever his precise relationship to Poseidonius may have been, it is generally agreed that his tactical treatise must be very closely related to that of Poseidonius. In fact the text of Asclepiodotus seems to be a précis of Poseidonius. It certainly lacks the introductory list of authors who had previously written works on the subject, ending with Poseidonius, which appear in Aelian and Appian. If this is the case, then it would be reasonable to suppose that the military organization described by Asclepiodotus should, at least in part, reflect the organization of the Seleucid Army in the final decades of the second century, albeit described through two filters; that of Poseidonius and that of Asclepiodotus himself. The probability is that Poseidonius is not so much describing the Seleucid Army, but rather using it as his principal source for terminology to supply the framework of an ideal army which could be divided down to unity. It would be reasonable to suppose that, in common with most other Greek philosophers, Poseidonius was interested in numbers, and hence his desire to demonstrate that the best way to organize an army, under ideal conditions, would be to divide it down to unity.

1923) p. 230; A.M. Devine, 'Polybius' lost *Tactica*: the ultimate source for the tactical manuals of Asclepiodotus, Aelian, and Arrian?' *AHB* 9.1 (1995) pp. 40-44.

[26] Poznanski, *op. cit.* ix; cf. Elizabeth Rawson, *Intellectual Life in the Late Roman Republic* (London 1985) p. 217 n. 14 'probably the pupil of Poseidonius'.

[27] Augustus Traversa, *Index Stoicorum Herculanensis* (1952) 73, 3.

[28] Thus Oldfather, *op. cit.* pp. 233-4; Poznanski, *op. cit.* pp. ix-xi. However Elizabeth Rawson, *Intellectual Life in the Late Roman Republic* (1985) p. 217 n. 14 identifies Asclepiodotus with the homonymous pupil of Panaitios.

In fact most Greek armies, at least at unit and sub-unit level in the infantry, were divisible to unity. This was perfectly understandable as the files of the infantry unit had to be able to halve and quarter in order to form close order and *pyknosis* ('locked shields'). Normally the infantry unit marched into battle with its files in open order, that is with each file occupying two yards of ground, which permitted it to move forward over rough ground with the minimum disruption to the formation of the files. When the unit came close to the enemy and these files would halve, with the back halves moving forward into the interstices of the front halves in order to form close order. Each file now occupied one yard of space. The general concept underlying the system is outlined in chapter four in Asclepiodotus (4. 1) [29]

> τριττὰ γὰρ ἐξηύρηται πρὸς τὰς τῶν πολεμίων χρείας, τό τε ἀραιότατον, καθ' ὃ ἀλλήλων ἀπέχουσι κατά τε μῆκος καὶ βάθος ἕκαστοι πήχεις τέσσαρας, καὶ τὸ πυκνότατον, καθ' ὃ συνησπικῶς ἕκαστος ἀπὸ τῶν ἄλλων πανταχόθεν διέστηκεν πηχυαῖον διάστημα, τό τε μέσον, ὃ καὶ πύκνωσιν ἐπονομάζουσιν, ᾧ διεστήκασι πάταχόθεν δύο πήχεις ἀπ' ἀλλήλων.
>
> The needs of warfare have brought forth three systems of intervals: the most open order (*araiotatos*), in which the men are spaced both in length and depth four cubits apart, the most compact (*pyknotatos*), in which with locked shields (*synaspikôs*) each man is a cubit distant on all sides from his comrades, and the intermediate (*messon*), also called a 'compact formation' (*pyknôsis*), in which they are distant two cubits from one another on all sides.

Asclepiodotus (4. 3) then proceeds to contradict what he has just said regarding the nomenclature of the three formations, but does supply us with information as to the tactical circumstances in which the closer spacings were used.

> τούτων δὲ τὸ μὲν δίπηχυ κατὰ πύκνωσιν, ἔφην, ἐπωνόμασται, τὸ δὲ πηχυαῖον κατὰ συνασπισμόν. γίνεται δὲ ἡ μὲν πύκνωσις, ὅταν ἡμεῖς τοῖς πολεμίοις τὴν φάλαγγα ἐπάγωμεν, ὁ δὲ συνασπισμός, ὅταν οἱ πολέμιοι ἡμῖν ἐπάγωνται.
>
> I have stated that of these two spacings the one of two cubits is called *pyknosis* and the one of a single cubit *synaspismos*. The former is used when we are marching the phalanx upon the enemy, the latter when the enemy is marching upon us.

---

[29] In this and the following quotation I have used the Loeb translation of Oldfather but given the Budé text of Poznanski.

Thus *synaspismos* was normally used only as a static formation, frequently to receive an enemy charge, though not always. In section 1a we have already mentioned that the file (*dekas*) in Alexander's army numbered 16, and that these were in turn formed into units numbering 512 and 1,024 men. The *lochos* of 512 men would normally be drawn up in *pyknosis* in 32 files of 16 men (i.e. a *dekas*) each. Thus it would occupy a frontage of 64 cubits, each file having a frontage of two cubits. To enable easy movement each even-numbered *dekas* numbering from the right would withdraw behind each odd-numbered *dekas*. Thus the frontage of the *lochos* would still be 64 cubits, but each file would now be 32 men deep and would occupy a frontage of four cubits. This most open order seems to have been called 'deep' (*eis bathos*) in Alexander's army, for this term is alluded to once by Arrian (*Anab.* 1. 2. 4). Alternatively the *lochos* could deploy from *pyknosis* into *synaspismos* should the tactical situation demand it. In this case the second-in-command of the *dekas,* stationed at the head of the rearward half of the *dekas*, would lead the second half-*dekas* of eight men into the interstice between the first half-*dekas* of his own *dekas* and the first half-*dekas* of the next *dekas* to the left. The *lochos* would still have a frontage of 64 cubits, but there would now be 64 men in the front line, each occupying a frontage of one cubit. The whole system in operation is described in a passage of the historian Callisthenes describing the march of the Macedonian phalanx into action at the battle of Issos preserved in Polybius (12. 19. 6)[30]

> ἅμα δὲ τῷ πρῶτον εἰς τὰς εὐρυχωρίας ἐκπεσεῖν, διασκευάζεσθαι παραγγείλαντα πᾶσιν ἐπιπαρεμβαλεῖν τὴν φάλαγγα, καὶ ποιῆσαι τὸ βάθος αὐτῆς ἐπὶ τριάκοντα καὶ δύο, μετὰ δὲ ταῦτα πάλιν εἰς ἑκκαίδεκα, τὸ δὲ τελευταῖον, ἐγγίζοντα τοῖς πολεμίοις, εἰς ὀκτώ.
>
> Immediately on issuing into the open country he re-formed his order, passing to all the word of command to form into phalanx, making it at first 32 deep, changing this subsequently to 16 deep, and finally as he approached the enemy to 8 deep.

Thus we can see the utility of a military organizational system dividing down into unity, and this system is documented in reality being used for the Alexandrine lochos of 512 men. (It must be stressed that whenever we talk about a file of 16, or any sub-unit of any other number, we are talking of an establishment strength, not a parade strength).

---

[30] Loeb translation and text of Paton. It is interesting that in his commentary immediately following his citation of this passage, Polybius choses to misunderstand the passage as implying that the frontage of the phalanx doubled then quadrupled.

At a higher level it would be natural to divide a formation into two so as to form two wings, or two halves of the main body of the centre. This would especially be true of the cavalry, which was normally required to provide cover to either flank of the main body of infantry in the centre. Armies increasingly started to operate in 'divisions' from the fourth century onwards: that is as they divided into parts in order to be able to advance over terrain more quickly, and only concentrated before battle. In this case too, the phalanx and all other arms would have to be divisible in order to supply each division with a contingent. Thus in general it was found that at a higher level it was useful to have formations organized along binary lines.

Thus it can be stated that the most convenient way to have a Graeco-Macedonian army organized was along binary lines. Needless to say, this is a general rule and not true of all armies at all times. The Hellenistic armies derived most of their body of tactical doctrine and organization from the practices of the Macedonian Army of Alexander from which they nearly all derived. The organization of Alexander's army formed the basis of all the Successor armies, including the Ptolemaic army, as we have already seen. Similarly, we know that the infantry of the Antigonid Army was organized into *speirai* of 256 men, and chiliarchies of 1,024 men.[31]

Modern armies tend to be organized in binary, triangular or 'square' patterns. The American army even tried out a 'pentagonal' scheme of organization in the post-War period.[32] To take a current example a British infantry section numbers 8 men, divided into two fire teams of 4, which is further divided into two pairs. The reason for this is to enable the infantry section to skirmish forward by 'fire and manoeuvre', one half supporting the other as it moves. Thus 'fire and manoeuvre' can be carried out down to the level of a pair. The most logical number for an infantry section is probably 7, as 6 is the highest number a human being can recognize without artificial systems of counting. Thus the section commander can instantly recognize if all his men are present or if someone is missing. The imperatives of tactical necessity do not, however, allow 7 to be used as the optimal size of an infantry section.

Higher up the organizational chain the number three starts to become prominent. There are three sections in a platoon, three platoons in a company, and three companies in a battalion and so on. This is because, whether in attack or in defence, it is felt desirable to deploy one's assets not just in two wings, but to hold back a

---

[31] Michel Feyel, *Rev. arch.* 1935, 2 p. 54; F. W. Walbank, *Philip V of Macedon* (Cambridge 1940) p. 293; Walbank, *Comm.* I p. 281.

[32] The pentomic infantry division was announced for the US army in 1957. It was to consist of five battle-groups of approximately 1,400 officers and men each. The pentomic organization was designed to facilitate dispersion in an atomic war scenario. The concept was abandoned in in 1961 as it was found to be too cumbersome and inflexible for the conditions imposed by the limited wars which were springing up everywhere. Cf *Infantry: Official Quarterly Publication of U.S. Army Infantry School,* vol. 47, no. 2 and 3 (April, July 1957).

reserve. The general scheme of deployment is therefore 'two up, one back'. A modern British armoured brigade is, however, generally 'square', in other words it consists of two armoured cavalry and two infantry regiments. This is, once again, for tactical reasons. The arms are equally balanced, but they will be divided up unequally, forming four 'battle-groups'. One battle-group may have two cavalry companies and one infantry company or vice-versa, depending on the tactical situation. The numbers two, three and four are not used in sequence to organize the British Army because the Chief of General Staff is an ardent student of the theory of numbers. Rather there are sound tactical reasons why these different numbers are used at different levels to organize the army. The fact that they are used in sequence is coincidental.

The concept of a reserve in battle was not well developed in Greek military thought. Battles were mainly fought in linear formations as the lack of a developed method of communication on the battlefield restricted the ability of the commander to react to unexpected developments on the battlefield. The job of a commander was to predict what would happen during the battle and to plan the deployment of his forces accordingly. Once the battle had started there was little he could do to change the deployment of his forces.[33] This is the reason why ancient Greek organizational systems tended to be binary from top to bottom.

We can only guess at Poseidonios' aim in writing the Τέχνη Τάκτικη. Perhaps it was to illustrate the inherent mathematical beauty, and universal tactical utility, of the binary system as a method of military organization. Whatever his aim was, there was certainly nothing outlandish in describing an army which was divisible down to unity, as for most of the Hellenistic period this was practically universally the case.

[33] On the development of Greek concepts of generalship see Everett L. Wheeler, 'The General as Hoplite' in ed. Victor Davies Hanson, *Hoplites: The Classical Greek Battle Experience* (London 1991) pp. 121-170.

APPENDIX C

The Date of the "Soldiers' Tomb" at Sidon

In 1897 a number of painted loculus slabs came to light near Saïda, the ancient Sidon, as a result of chance discovery. Some were moved into the caravan-serai in Saïda the next year. The discoveries had been made in the garden known as Bostan el-Hamoud to the south of the town, at the foot of the hill dominated by the ancient fortress. A few of the loculus slabs discovered in 1897 were left where they had been found. The discoveries had aroused quite a substantial amount of curiosity in the scholarly world, and in 1903 Macridy Bey carried out a sondage in the area on behalf of the then Imperial Ottoman Archaeological Museum in order to recover what he could of the material which still remained. Three pits were sunk into the garden. From the first came numerous fragments of stucco belonging to loculus slabs destroyed in the course of the earlier haphazard excavations, and from the other two a diverse selection of material, including some late Hellenistic pottery and two fragments of vases decorated in relief with representations of the god Bes. At a depth of seven metres a wall was found, late in date, constructed of re-used material including a number of painted loculus slabs. In total seventeen complete or fragmentary loculus slabs from these various excavations have been published, but of these only seven survive, thanks to the efforts of Macridy Bey, who transported them to the Archaeological Museum, Istanbul.

All the loculus slabs showed deceased warriors. Macridy Bey concluded that the material originally came from a military necropolis established by foreign mercenaries in the vicinity. In context, style and format the loculus slabs obviously all belong to a coherent group executed by the same artist or workshop at around the same time. They seem to come from one or more tombs of the 'hypogeum' type, with lateral niches or 'loculi' for individual burials. The loculus slabs were used to seal an individual loculus in the tomb. They are all in the shape of funerary stelai, and so are sometimes referred to as stelai in the literature. Consequently the group of material as a whole is frequently referred to as the 'Soldiers' Tomb' from Sidon, on the assumption that the material came from a single hypogeum.[1] It is reasonably safe to conclude that these mercenaries all belonged to units comprising a garrison

---

[1] For other painted loculus slabs from Saïda, recovered at an earlier date and coming from other hypogea, see Ch. Clermont-Ganneau, 'Stèles peintes de Sidon' *Gazette archéologique* 3 (1877) pp. 102-115 pls. 15-16; Lammens, *Rev. arch.* 1898, 2 p. 109.

stationed in the ancient city of Sidon.

This group of archaeological material has not been subjected to the scholarly attention it deserves. Academic opinion is divided as to whether the troops belonged to the Seleucid or Ptolemaic army, and consequently as to the date of the material. It has been thought that if the loculus slabs are Ptolemaic, they should date to the third century, as Koile-Syria was lost to the Seleucids in the 190s BC. Consequently, it is argued, if the loculus slabs are Seleucid they should belong to the second century. The current majority opinion seems to be that they are Ptolemaic, dating to the late third century, though a minority opinion holds that they are Seleucid dating to the second.[2] The fact that the soldiers named on the loculus slabs come from Greece and Anatolia seems to rule out the second possibility, for the Seleucid monarchs were banned from recruiting west of the Taurus Mountains by the terms of the Treaty of Apameia. A third possibility exists; that the tombstones are Ptolemaic, but date to the second century. This could be possible if they had been deposited during one of the military intervention(s) of Ptolemy VI Philometor in Syria during the years 150-145 BC. The purpose of the following discussion is to summarize the arguments in support of this third possibility.

Section C 1

The Military Intervention of Philometor in Syria

During the second century the Seleucid royal house was divided into two branches. In 175 Seleucus IV was murdered and his brother Antiochus V Epiphanes took the throne. The previous year Seleucus' son Demetrius had been sent to Rome as a hostage. When Antiochus V Epiphanes died in 164, rule passed to his infant son Antiochus V Eupator and his highly unpopular regent Lysias. In 162 Demetrius decided to make an attempt at the throne, Antiochus V and Lysias were murdered, and Demetrius assumed rule as Demetrius I.

During the 150s Demetrius I of Syria made himself increasingly unpopular and surrounded himself with enemies both within and without the Seleucid Empire.

---

[2] Thus Mary Hamilton Swindler, *Ancient Painting from the Earliest Times to the Period of Christian Art* (New Haven 1929) p. 346 thought they were Seleucid dating to the second century, whilst E. Bikerman, *Institutions des Séleucides* (Paris 1938) pp. 88-90 doubted that they were Seleucid, as did M. Rostovtzeff, *The Social and Economic History of the Hellenistic World2* (Oxford 1953) III p. 1401 n. 137 'more probably Ptolemaic than Seleucid'. See also *ibid.* p. 150 'Late third or early second century B.C.'; Bar-Kochva, *Judas Maccabeus* p. 581 pl. IX 'Ptolemaic ... towards the end of the third century B.C.'

Consequently Attalus of Pergamon and Ptolemy VI Philometor of Egypt recognized as king a young man residing at Smyrna called Alexander Balas, who claimed to be a son of Antiochus Epiphanes. In 152 Balas landed at Akê-Ptolemaïs. Demetrius withdrew his forces from Judaea in order to concentrate them against Balas, and granted fairly complete independence to the high-priest Jonathan. Jonathan accepted all these privileges, but then switched his support to Balas. It may be that Philometor sent an expeditionary force into Syria in support of Balas, for Diodorus (33. 20. 1) tells us that 'in the war against Demetrius', 'those of the army from Alexandria' were commanded by Galaistes, the son of Amynander, the last king of the Athamanians. The two armies met, probably in the summer of 150, and though the army of Demetrius was victorious, Demetrius was killed. Balas established himself in Syria and proposed to Philometor that the two kingdoms join in an alliance. They met at Ptolemaïs where Balas was married to Philometor's daughter Cleopatra.

Over the next couple of years Balas proved himself to be a weak monarch and rule was actually exercised by one Ammonios. Apollonius, Alexander's governor of Koile-Syria, was unable to keep Jonathan and his Jewish army in check. They defeated Apollonius in battle, and then attacked the city of Azotos, burning the temple and eight thousand who had taken refuge within it. Disaffection with the new regime was widespread, and in 147 the teenage son of Demetrius I appeared in Cilicia laying claim to the throne. Philometor, together with an army and a fleet, entered Koile-Syria to fight as an ally of Balas, passing through Azotos along the Palestinian coast as far as Ptolemaïs. All the cities welcomed Philometor as their defender against the excesses of Jonathan, and Philometor left a garrison of troops in each town he passed through (*1 Macc.* 11. 3 'As he went on his progress from town to town, Ptolemy left a detachment of troops in each of them as a garrison'). At Ptolemaïs Philometor met Balas, and Ammonios unsuccessfully tried to have Philometor assassinated. When Balas refused to hand over Ammonios for punishment, Philometor transferred the hand of his daughter, together with his military support, from Balas to Demetrius II. The Egyptian intervention in Syria lasted until 145 when Philometor met his death at the battle of the River Oinoparas. The Ptolemaic army in Syria and Koile-Syria was left leaderless and rapidly collapsed. Some units were incorporated into the Seleucid army, while others left the country.

In conclusion, therefore, there is some slim evidence for a first military intervention of Philometor in support of Balas against Demetrius in 150 BC, in the form of the text of Diodorus 33. 20. 1, but this intervention was of a temporary nature, and we have no evidence that a Ptolemaic garrison was installed in Sidon. For the second intervention, that of 147-145 BC, we have the explicit statement of *1 Maccabees* (11. 3), that Philometor left a garrison of troops in each (Levantine) town he passed through. It needs to be stressed that this is the only evidence for a Ptolemaic military presence in the Levant in general, and in Sidon in particular, during the whole of the second century down until the conflict of 103-101 BC, known as the 'War of Sceptres'.[3]

We shall now consider some general chronological considerations with regard to the loculus slabs from Sidon as a whole.

Section C 2

Stylistic Considerations

Stylistic considerations can rarely be used to assign dates to ancient archaeological material with a margin of error of anything less than several decades. Nevertheless, when art historians have dealt with the Sidon loculus-slabs, they have tended to give them a later, rather than an earlier date. Blanche R. Brown thought that the loculus-slabs were all in the 'popular' style which only emerged in the second century. The 'popular' style marked a drop from the dominant level of artistic production 'to a low, popular level which is characterized by simplified, conventionalized forms and compositions'. She noted that, although the Sidon loculus slabs derive from a different stylistic source than the Alexandrian material which she was studying, and they therefore use different decorative motifs, they are nevertheless comparable to the tombstones belonging to her 'Fourth Style', which date to the second century. They 'isolate, abstract, freeze, and repeat symbols', a phenomenon which can also been seen in the treatment of the human figures of the Sidon material.[4]

Peter Callaghan[5] considered that the Trefoil Style wreath, which hangs in the frieze below the pediment on many of these loculus slabs, indicates a date in the 'latter half of the period of Seleucid domination' (i.e. the second half of the second century). He realized that this date caused historical problems, as many of the mercenaries came from west of the Taurus Mountains, and so should not have been recruited for service in the Seleucid army according to the terms of the Treaty of Apamea, but affirmed his conviction that 'the style of the stelai demand that they be placed in this latter period'. If the soldiers were in Ptolemaic rather than Seleu-

---

[3] E. Van't Dack, W. Clarysse, G. Cohen, J. Quaegebeur & J. K. Winnicki, *The Judean-Syrian-Egyptian Conflict of 103-101 B.C.. A Multilingual Dossier Concerning a "War of Sceptres" (Collectanea Hellenistica I,* Louvain 1989) passim.

[4] *Ptolemaic Paintings and Mosaics and the Alexandrian Style* (Cambridge Mass. 1957) p. 46 'the painted loculus slabs of Sidon, made for soldiers of the Seleucid garrison some time after 195 B.C.' p. 87; cf. Mary Hamilton Swindler, *Ancient Painting from the Earliest Times to the Period of Chistian Art* (New Haven 1929) p. 346 'They are cruder in style and less Greek ... They date in the Second Century and in later times'.

[5] In *BSA* 75 (1980) p. 45.

cid service, of course, this problem disappears.

The letter-forms of the inscriptions could be put in the late third century, but would fit a date in the middle of the second century well. The letters are not apicated: that is, the legs do not splay out at the ends, but there is some thickening. The broken-barred *alpha* appears, a feature which is rarely found before the second century. Likewise the top and bottom strokes of the *sigma* are parallel, which is another feature characteristic of the second century and beyond. The *theta* has a stroke in the middle, and this feature is rarely found before the middle of the second century. On the other hand there is some fine curving in the letters *upsilon* and *alpha*, which is not a characteristic feature of a date any later than the second, or even the third, century. The *pi* has its right leg shorter than the left in most cases, but in some the legs are equal, and this lengthening is a process which took place during the second century. In some cases the *omicron* is equal to the other letters in size, but in others it is small and hangs in the middle of the line, which is a feature which disappears in the late third century. For these reasons a date in the middle of the second century would fit the epigraphic style well. Arguments based purely on letter-forms can never, however, be regarded as conclusive.

Section C 3

The *Politeuma*

The funerary inscriptions of many of the soldiers tell us that they belonged to *politeumata*. *Politeumata* are thus far only attested for the Ptolemaic army, and only for the second century BC. It is, indeed, possible that they were an innovation dating to the reign of Philometor. *Politeuma* could be translated as 'citizen body', but how the system actually functioned in Egypt is more or less completely unknown.[6] Although the *politeumata* are thus far attested as existing only in Ptolemaic Egypt, this may simply be a reflection of the state of the evidence. Papyrological evidence, of which a substantial proportion is concerned with legal matters such as an individual's membership of a *politeuma*, has only survived for Egypt in any quantity, but similar evidence has not survived for the Seleucid or Antigonid Empires. Consequently it would be extremely hazardous to assert that the *politeumata* were an exclusively Ptolemaic institution, though the evidence suggests this at the moment.

---

[6] On the *politeumata* at Sidon see M. Rostovtzeff, *The Social and Economic History of the Hellenistic World III2* (Oxford 1953) p. 1401 n. 137; Launey pp. 1081-1084.

The Ptolemaic *politeumata* seem to have been created at a specific date for a specific purpose, and they are named after different ethnic groups: 'the *politeuma* of the Cretans' for example. If *politeuma* did exist as a word or as an institution in the Seleucid Empire, for which there is as yet no firm evidence, there is no reason why it should have been in exactly the same form as the Ptolemaic *politeuma*. All that can be said is that the *politeuma*, where it is attested in the inscriptions on the loculus slabs from Sidon, is an ethnic group of exactly the same type as the Ptolemaic *politeuma*. A Ptolemaic interpretation would therefore fit this material very well.

Many scholars[7] have simply assumed the *politeumata* to have been an institution common to all the Hellenistic kingdoms. Tarn[8] has postulated the existence of a *politeuma* of Syrians in Seleuceia-on-Tigris from a passage in Josephus (*AJ* 18. 372), who tells us that during the Parthian period there lived in Seleuceia many Macedonians, even more Greeks, and there were also not a few Syrians 'enrolled in the citizen body' (οἰκοῦσιν δ' αὐτὴν πολλοὶ μὲν Μακεδόνων, πλεῖστοι δὲ Ἕλληνες, ἔστιν δὲ καὶ Σύρων οὐκ ὀλίγον τὸ ἐμπολιτευόμενον). Tarn argued that this last word 'which is the verb of πολίτευμα, not of πόλις or πολίτης' demonstrated that the word *politeuma* was not confined to Egypt. But ἐμπολιτεύω it is not the verb of *politeuma*; rather it means 'to hold citizen rights [in]', from the simple form of which the noun *politeuma* is derived. The verb is found in sources as early as Thucydides (4. 103, 106). Elsewhere Josephus (*AJ* 18. 378) tells us that whoever of the Syrians 'who was a citizen' (ὁπόσον ἦν Σύρων ἐμπολιτεῦον) joined in the hostility to the Babylonian Jews. What these passages seem to imply is that Seleuceia, and the other cities which had once been within the Seleucid Empire, had a single citizen body, which may or may not have been called a *politeuma*, which included many Macedonians, more Greeks, and not a few Syrians. The passages do not seem to imply that Seleuceia had a *politeuma* of Macedonians, a separate *politeuma* of Greeks and another *politeuma* of Syrians, which is what Tarn wished to read into the passage.

An *argumentum e silentio* is, however, never secure. We might compare the Ptolemaic institution of the *epigonê*. The bearers of this status seem to be descendants of the original Graeco-Macedonian settlers who constituted the Ptolemaic body-politic.[9] The institution is attested in innumerable papyri from Egypt, but outside Egypt only in a single funerary inscription (assuming the reading to be correct) from Pagasai-Demetrias,[10] commemorating 'Sotion son of Dionysios, one of the *epigonoi*'.

---

[7] Mørkholm, *Antiochus* p. 138 n.12.

[8] W. W. Tarn, *The Greeks in Bactria and India*^{2} (Cambridge 1951) p. 18 nt. 5; accepted by G. A. Downey, *A History of Antioch in Syria from Seleucus to the Arab Conquest* (Princeton 1961) p. 115.

[9] Lesquier p. 57

[10] Arvanitopoullos in *Polemon* 4 (1949/50) pp. 83-4 no. 256.

Σωτίων
Διονυσίου
τῶν ἐπιγόνων

No ethnic is given in this inscription, which, presumably, should be taken as an indication that the deceased was a citizen of Demetrias. What the institution of the *epigonoi* was in the Antigonid kingdom, and what relationship it bore to the Ptolemaic institution of the *epigonê* are unknown, but had this single inscription not survived, we might have assumed that the Ptolemaic institution of the *epigonê* was an isolated phenomenon.

While keeping all these doubts in mind, it remains true that up till now evidence for *politeumata* of the type encountered in the inscriptions on the Sidon stelai is restricted to Egypt. A Ptolemaic context is therefore, on face value, more probable than a Seleucid one.

Section C 4

The Office of *Hêgemon ep'Andrôn*

One stele from Sidon may provide further evidence that the group is Ptolemaic rather than Seleucid, but, as is the case with the institution of the *politeuma*, this also depends on an *argumentum e silentio*. The stele was heavily damaged when found and has not survived, consequently its reading and interpretation are open to dispute. The stele was first recorded in 1903 and published in 1904 by Macridy Bey and Jalabert separately.[11] An autograph copy was made by Macridy Bey, which is given below, together with his restored text.

[. . .Ἑρ]μολύκου
[. .]νθινη. ε Μο[σ]
χίων [χρ]ηστὲ
χαῖρε.

This minimalist restoration is clearly dubious. Working independently, Jalabert published his own reading and restoration. He believed the right side of the inscription was complete, while the left side lacked four or five letters.

[11] Jalabert *Rev. arch.* 1904, 2 pp. 3-4 no. 1; Macridy Bey, *Revue Biblique* 1904, p. 554 no. 9.

Figure 32. Autograph copy made of the funerary inscription of the son of Hermolykos by Macridy Bey (after *Rev. Bibl.* 1904, p. 554).

[. . . . . α Ἑρ]μολ[ύ]κου
[. . . .]ΝΦΙΝ ἡ[γ]εμό[να]
[. . . . .]ΧΙΩΝ [χρ]ηστ[έ]
χαῖρε.

Jalabert noted that while the patronymic Hermolykos, a very common personal name, is beyond doubt, the fragmentary ending ΧΙΩΝ in line 3 could be supplemented in two ways. Firstly one could restore ἡ[γ]εμό[να] | [τῶν] Χίων which would make the deceased an 'officer of the Chians', but Jalabert noted that this would be both dubious and contrary to normal usage. More probably, Jalabert thought, the first name of the deceased, perhaps Moschion, reappeared in line 3. He finally noted that it is difficult to read any ethnic, which would normally occupy that position, in line 2. Thus the inscription might read as follows.

[Μοσχίωνα Ἑρ]μολ[ύ]κου
[. . . .]ΝΦΙΝ ἡ[γ]εμό[να]
[Μοσ]χίων [χρ]ηστ[έ]
χαῖρε.

[To Moschio]n son of Hermolykos
of [. . . .]nthos, an officer.
[Mos]chion! Good man!
farewell!

Launey also attempted a restoration.[12] The letters NΘIN or NΦIN in line 2, he suggested, must form part of the ethnic of the deceased. He suggested that line 3 had been read and restored incorrectly by both Jalabert and Macridy Bey. He proposed a restoration of ἡ[γ]εμό[να] | [ἐπ' ἀν]δρῶν (*hêgemon ep'andrôn*). The exact meaning of this title of rank is unclear,[13] but it was perhaps given to a unit commander as opposed to the term of *hêgemôn* 'officer' in general. Thus, if Launey's restoration is correct, it gives a general indication that the Sidon tombstones are Ptolemaic. Nevertheless, the title may also have been in use in the Seleucid army too, but simply not recorded. Furthermore, while the restoration is logically a most attractive one, it cannot be reconciled easily with the autograph copy of the damaged lettering reproduced by Macridy Bey. It would require that both Macridy Bey and Jalabert incorrectly read an elongated *delta* as a *chi*, and the vertical stroke of the *rho* as an *iota*.

Section C 5

The Chronological Evidence Given by the Ethnics

Of the 17 loculus slabs from Sidon, three are anepigraphic, and one other does not record an ethnic for the deceased. Of the remaining 14 inscriptions four record deceased soldiers from the Greek mainland and islands, while the remaining 10 deceased came from different areas of Anatolia, 1 from Lydia, 2 from Lycia, 2 or 3 from Caria and 3 or 4 from Pisidia. Thus two thirds of the deceased were recruited from Anatolia north-west of the Taurus Mountains. We shall return to this point later on.

One of the loculus slabs, now in Istanbul Archaeological Museum (Inv. 1489), was established in memory of a Pisidian, Saëttas son of Trokondas 'one of the Termessians of those near Oinoanda.'[14] The inscription describes Saëttas as a *symmachos* from the Termessians. It seems that the terms 'Termessos by Oinoanda' and 'Oinoanda' refer to the same city. Originally Oinoanda may be have been either the name of a geographical location, or perhaps the name of a scattered community, which became settled by colonists from Termessos Major during the Hellenistic period.[15] Termessos by Oinoanda was also known as Termessos Minor.

---

[12] Launey p. 615 n.6.

[13] Lesquier p. 85-7.

[14] Gustave Mendel, *Musées Impériaux Ottomans. Catalogue des sculptures grecques, romaines et byzantines* 1 (Istanbul 1912) pp. 262-3 no. 103; Sayce, *Classical Review* 28 (1914) 197 no. 4.

[15] J. J. Coulton, 'Termessians at Oinoanda' *AS* 32 (1982) pp. 115-131, a position accepted by M. Wörrle in *Stadt und Fest im kaiserzeitlichen Kleinasien* (Vestigia vol. 39, 1988) pp. 45-53.

Theoretically, therefore, this inscription should give us a *terminus* date for the group of material as a whole. The date of the foundation of the colony cannot, however, be securely established. The evidence suggests a date between 225 and 175 BC,[16] but attempts to tie the date down further have not been successful.[17]

The ethnic used in a further loculus slab, however, that of Aristeidas son of Aristeidas 'a Lakedaimonian from Gytheion' supplies us with slightly more certain chronological evidence.[18] Jalabert[19] argued that in about 195 BC, the Roman general Flamininus forced the tyrant Nabis of Sparta to recognize the independence of the Lakonian cities, including Gytheion, which had formerly been under Spartan domination. These were, seemingly, organized into a 'League of the Lakonians' which was placed under the protection of the Achaean League. Consequently, Jalabert continued, the ethnic 'Lakedaimonian from Gytheion' could not have been used before 195. Two objections could be raised to Jalabert's argument. In the first place it is not certain that the League was created as early as 195, an alternative date might be 146 following the dissolution of the Achaean League.[20] Secondly Bikerman stated very firmly that there is absolutely no reason why the ethnic 'Lakedaimonian from Gytheion' may not have been in use before the establishment of the League of the Lakonians.[21] The sparse evidence tends to support Jalabert however, for only a second example exists of an ethnic of this type. The athlete Nikokles son of Nikatas used the ethnic 'a Lakedaimonian from Akriai' in an inscription dateable to the first half of the first century BC.[22] Like Gytheion Akriai was also one of the Eleutherolakonian cities.[23] The ethnic one would expect to be used by an inhabitant of Gytheion before the establishment of the League is 'Gytheian' (Γυθεάτης) without further qualification. The earliest firmly dated example being, indeed, used on an honorific statue for Titus Flamininus in 195 BC.[24] Consequently, although certainty still eludes us, there may well be some substance in Jalabert's suggestion that the ethnic used by Aristeidas gives a date later than 195 to the material from Sidon.

The loculus slab of the Perrhaibian Eunostides son of Nikanor also survives in Istanbul.[25] The reading of Eunostides is not secure. Jalabert saw the inscription

---

[16] Coulton, *op. cit.* 129-130.

[17] Cf. L. Robert, *Hellenica* 10 (Paris 1953) p. 196 n. 3; E. Bikerman, *Institutions des Séleucides* (Paris 1938) p. 89 n. 1.

[18] Macridy Bey, *Revue Biblique* 1904 pp. 552-3 no. 5; *Rev. arch.* 1904, 2 pp. 8-9 no. 5.

[19] E. Jalabert, *Rev. arch.* 1904, 2 pp. 8-15.

[20] Paul Cartledge & Antony Spawforth, *Hellenistic and Roman Sparta. A Tale of Two Cities* (London 1989) p. 77, cf. p. 100.

[21] E. Bikerman, *Institutions des Séleucides* (Paris 1938) p. 89 n. 1.

[22] *LGPN* III A *sv.* I would like to thank the *Lexicon of Greek Personal Names*, through the offices of Richard Catling, for the information on ethnics given in this and the following couple of notes.

[23] Paus. 3. 21. 7.

[24] *IG* V (1) 1165.

[25] Gustave Mendel, *Musées Impériaux Ottomans. Catalogue des sculptures grecques, romaines et*

first, and restored Eunostides rather than Eunostos without hesitation, because he saw a vertical *hasta* to the right of the *tau*. Macridy Bey, however, restored Eunostos (in the genitive) which does, it must be said, fit the available space better. He also read the ethnic in the genitive, which is clear in his autograph copy of the inscription. Jalabert's reading has generally been accepted. Stählin first suggested that our Eunostides son of Nikanor buried in Sidon was probably the son of Nikanor son of Eunostos of Gonnoi, mentioned in an inscription dating to around the second half of the third century.[26] Gonnoi was one of the principal cities of Perrhaibia, an outlying district of Thessaly lying on the Macedonian border. Eunostos is a personal name of Boeotian origin which had presumably spread though close personal contact between important Perrhaibian and Boeotian families. It is found in another inscription from Gonnoi, but nowhere else in Perrhaibia. It would be reasonable to conclude, therefore, that Eunostides did in fact belong to this significant family from Gonnoi, though the precise family relationship between him and the earlier Eunostos must remain uncertain.

Eunostides uses the ethnic 'Perrhaibian' on his tombstone, rather than 'Gonneus'. At first sight this strikes one as being peculiar. Although we have too little epigraphic evidence to be absolutely certain on the matter, it seems that the ethnic 'Perrhaibian' was only used for the short period of time during which the Perrhaibian League was in existence.[27] The Perrhaibian League was established by a declaration of Flamininus made during the Isthmian Games of 196 BC (Livy 33. 32. 5), and existed down to 146 BC or shortly after, when it was annexed to the Thessalian League in the wake of the 'Revolt of Andriskos'; the Fourth Macedonian War.[28] We know that the revolt spread into Thessaly. Even though we have no information as to how it affected the cities of the Perrhaibian League, it is highly probable that the cavalry of the League, to which Eunostides belonged, remained loyal to Rome. The final military defeat of Andriskos in 148 would have freed the cavalry of the Perrhaibian League to enter Ptolemaic service. They could, therefore, have joined the Ptolemaic army in time to participate in the second intervention of 147-145.

Presumably the Perrhaibian troops would have been sent in accordance with a treaty of alliance pledging reciprocal military aid in time of war. There is no evidence that Philometor had sent any troops to the aid of the Perrhaibians during the Revolt. He may, however, have sent out military equipment and perhaps money in lieu of manpower to help them. By way of comparison we might mention that for example in 188 BC his father Ptolemy V Epiphanes had exchanged oaths of alli-

---

byzantines I (Istanbul 1912) pp. 269-270 no. 108.

[26] Friedrich Stählin, *Das hellenische Thessalien* (Stuttgart 1924) 8 n. 3; Bruno Helly, *Gonnoi II* (Amsterdam 1973) no. 232

[27] Cf. Launey p. 218 n. 1.

[28] Cf. H.Kramolisch in B.Helly (ed.), *La Thessalie, Actes de la Table Ronde 21-24 juillet 1974* (Lyon 1979) pp. 210-11.

ance with an embassy, which included Lykortas father of Polybius, sent by the Achaean League to Alexandria, and had sent the League 6,000 bronze shields for peltasts and 200 talents of coined gold (Polyb. 22. 9. 3).

It is possible that Eunostides travelled to Egypt as an individual to fight as a mercenary in the Ptolemaic army, in which case the *terminus* dates supplied by the stele would be as wide as 196 - *c.* 148. However Eunostides' groom holds his weapons, a pair of spears and a square shield in the shape of a conventional *thureos* shield with the top and bottom cut off. These are the weapons of a 'Tarentine' cavalryman. We know from epigraphic evidence that the cavalry of the Thessalian League was commanded by a hipparch and a tarantinarch.[29] It seems to me on balance to be more likely that we should take this stele as an indication that the cavalry of the Perrhaibian League included a regiment of 'Tarentines', and that this regiment was sent '*en bloc*' to Egypt shortly after 148, rather than as an indication that the Ptolemaic army contained a regiment of Tarentines which Eunostides joined as an individual. Certainty once again escapes us in this matter, but what is important is that the stele of Eunostides supplies us with a reasonably firm *terminus post quem* of 196 BC and a *terminus ante quem* of shortly after 146 BC.

Section C 6

The Stele of Dioskourides

Perhaps the firmest evidence as to the chronology of the Sidon loculus slabs is provided by that of Dioskourides of Balboura. The loculus slab is preserved in the Istanbul Archaeological Museum (Inv. 1490).[30] The inscription runs as follows:-

Διοσκουρίδη 'Εξαβόου Πισίδη
Βαρβουλεῦ συμμάχων
σημεοφόρε χρηστὲ

[29] Luigi Moretti, *Iscrizioni Storiche Ellenistiche II* (Florence 1975) 130; *IG* ix (2) 509.

[30] Gustave Mendel, *Musées Impériaux Ottomans. Catalogue des sculptures grecques,romaines et byzantines* I (Istanbul 1912) pp. 259-61 no. 102 (with previous literature); A. H. Sayce, *Classical Review* 28 (1914) p. 197 n. 3; Paul Couissin, *Les Institutions militaires et navales* (Paris 1932) pl. 38, 2 (drawing); Blanche R. Brown, *Ptolemaic Paintings and Mosaics and the Alexandrian Style* (Cambridge Mass. 1957) p. 87, pl. 20, 2; Volkmar Von Graeve, *Der Alexander Sarkophag und seine Werkstatt* (= *Istanbuler Forschungen 28,* Berlin 1970) 89 n. 47; Petros Dintsis, *Hellenistische Helme* (Rome 1986) pl. 41, 3; Klaus Parlasca, *Syrische Grabreliefs hellenistischer und römischer Zeit. Fundgruppen und Probleme* (= *Trier Winckelmannsprogramme 3,* 1981) pp. 6, 23 n. 14, pl. 2,3; Bar-Kochva, *Judas Maccabeus* p. 582 pl. x.

χαῖρε
Κεραίᾶ ὁ ἀδελφός ἔστησε.

Oh Dioskourides son of Exaboos, a Pisidian,
a Balbouran of the *symmachoi*,
standard-bearer. Good
man farewell!
Keraias his brother set up (this stele).

The place name occurs here in the form Barboula, which is not unusual and occurs frequently in later attestations of the city. Sayce mis-interpreted the patronymic as a second place-name 'from Aboos', but Exaboas seems to be a local name and is listed as such by Zgusta.[31] Robert[32] was uncertain whether the personal name of the brother of Dioskourides, Keraias, should be regarded as Pisidian because, although it is found in a number of other inscriptions from the Kibyratis, it is also found in the Greek city of Colophon. Zgusta (580-2) considered the name to be certainly Anatolian.[33]

The title of 'standard-bearer' given to Dioskourides places the stele firmly in the second century. This title of military rank is not found in any Hellenistic army before the 160's, and, as we have already seen in the main body of the text above, its début is associated with the introduction of military equipment and organization along Roman lines into both the Ptolemaic and Seleucid armies in the 170's and 160's.

The spelling of Dioskourides' rank of standard-bearer, σημεοφόρος, is possibly of some chronological significance too. The standard spelling in Ptolemaic documents is σημεοφόρος. The variant spelling, as on the stele of Dioskourides, occurs in the earliest dateable occurrence of the rank in Ptolemaic documents, which has already been mentioned. Korax son of Dionysios is named as a σημεοφόρος in the regiment of Pasines at Hermopolis Magna in 143 BC (*PP* 2385). Other occurrences of this form of spelling are not dated, and we have an insufficient sample of material to enable us to decide whether we are dealing with a shift in spelling over time, or with variant forms of spelling without chronological significance. Nevertheless it is possible to regard the spelling of Dioskourides' rank as a possible indication of a mid-second century date.

Curiously, Dioskourides does not carry a standard (see Figure 15), and one wonders whether the Hellenistic armies had simply taken over the name of the rank from the Roman army, but not the practice of carrying standards. We might

[31] L. Zgusta, *Kleinasiatische Personnennamen* (Prague 1964) p. 340-1.

[32] Louis Robert, *Études anatoliennes* (= *Études orientales publiées par l'Institut Français d' Archéologie de Stamboul 5,* Paris 1937) pp. 366-7.

[33] L. Zgusta, *Kleinasiatische Personnennamen* (Prague 1964) pp. 580-2.

compare the word ensign in modern usage. There might be some evidence for standards physically being carried in a funerary epigram which has been recovered from Egypt.[34] One Ptolemaios had served in a Macedonian regiment both as a standard-bearer and as a *hêgemôn* at various times. He had fought as a bold warrior or spearman 'with his standard-bearer's staff' (σημοφόρωι κάμακι).

Section C 7

Conclusions

As far as the Sidon loculus slabs as a whole are concerned, the argument for a *terminus post quem* given by the occurrence of the city name of Termessos by Oinoanda as an ethnic is confused and inconclusive. Jalabert's argument concerning a possible reference to the League of the Lakonians, while likewise not conclusive, is more persuasive, and if correct should give a *terminus post quem* of 195 BC to the group as a whole. The occurrence of the ethnic Perrhaibian, if my arguments connecting it with the Perrhaibian League are accepted, should supply *terminus* dates of 196 BC to 146 BC or shortly after for the group as a whole. Thus a second century date for the Sidon stelai is suggested by the evidence of the ethnics.

Perhaps the firmest indication of the date of this group of tombstones is the rank of standard-bearer which Dioskourides of Balboura holds. This rank only existed in either the Seleucid or Ptolemaic army after the 170's and 160's, which, in my view, gives the firmest *terminus post quem* to this group of material. Likewise, the 'Romanized' military equipment worn by Salmas, discussed in the main body of the text, argues for the same *terminus post quem*.[35] A date no earlier than the 160s seems inescapable. So the *terminus* dates are thereby narrowed to *c.* 170- *c.* 146 BC.

During this period the areas of Anatolia from which the majority of the troops buried at Sidon came were, if not independent, subject to Attalid inflence. It is difficult to see under what circumstances the Seleucid monarchs could have recruited these troops. For much of this period, following the death of Antiochus V, the Attalid and Seleucid kings were at daggers drawn. The terms of the Treaty of Apamea forbidding recruitment of mercenaries west of the Taurus might be bent, but this only happened occasionally and under particular circumstances. Following the Treaty of Apamea the Seleucid monarchs would certainly not have been able to

---

[34] Étienne Bernard, *Inscriptions métriques de l'Égypte gréco-romaine* (Paris 1969) p. 50.

[35] Petros Dintsis, *Hellenistische Helme* (Rome 1986) I p. 290 no, 248 dated this stele to the first half of the second century.

conclude symmachic military alliances with the Pisidian cities, which are attested in the inscriptions on the tombstones of Dioskourides and of the Pisidian Saëttas 'one of the Termessians of those near Oinoanda' discussed in the text above. Let us recall that of the fourteen inscriptions where the ethnic is preserved, no less than ten record that the deceased came from areas of Anatolia north-west of the Taurus. It is therefore unlikely that the tombstones were deposited by troops in Seleucid service.

If a Ptolemaic interpretation were to be accepted this difficulty would disappear, and the appearance of *politeumata* and the rank of *hêgemôn ep'andrôn* in the inscriptions would cause no difficulties. If the troops were Ptolemaic, then only two dates are possible within the stated range: the dubious intervention of a Ptolemaic army commanded by Galaistes in support of Alexander Balas in 150, and the second intervention of Philometor in the years 147-145. We are not told that garrisons were left in any of the cities which the army passed through in 150, but we are specifically told (*1 Macc.* 11. 3) that garrisons were left in the cities by Philometor during the intervention of 147-145. For this reason the latter date is the one to be preferred, and this would allow sufficient time for the Perrhaibian allied cavalry to be sent following the defeat of Andronikos.

The ethnic 'Perrhaibian' is the principal argument suggesting that the Sidon tombstones should not be taken down in date later than *c.* 145 BC, but it may also be worth recalling the stylistic considerations discussed above which likewise suggest that a date later than the second century, and perhaps even a date later than the middle of the second century, is unlikely. For these reasons it is highly unlikely that the Sidon material can be dated as late as the Ptolemaic intervention of 103-101 BC.

The Ptolemies had particularly close relations with the Lycian League: this is why we find so many Lycians serving in the Cyprus garrison and in the Sidon garrison too. Lydians are also mentioned in the inscriptions. Lydia was, however, incorporated within the Attalid Kingdom at this period, and we may perhaps conclude that Philometor was allowed to recruit there during the period when the Ptolemaic and Attalid kingdoms had entered into a coalition to dethrone Demetrius I and replace him with Alexander Balas. Symmachic alliances between Philometor and the Pisidian cities of Balboura and Termessos by Oinoanda are attested in the funerary epitaphs of Dioskourides and Saëttas. It may be the case that Philometor had also contracted similar alliances with the Perrhaibian and Lakonian Leagues. If so, this suggests that Ptolemaic diplomatic activity in Greece during this period was more widespread than has previously been suspected.

APPENDIX D

The Purpose and Date of the Daphne Parade

In 167 the Roman general Aemilius Paulus had celebrated his victory at Pydna with Games held at Amphipolis. The text as it stands in Athenaeus (5. 194 C sq.) tells us that Antiochus, hearing about the Games held by Aemilius Paulus, resolved to outdo him with even more magnificent Games to be held near the Syrian city of Daphne the next year. This explanation is accepted at face value by most modern historians.[1] It is possible, however, that the stated reason did not appear in the original text of Polybius from which this passage derives, and could be a speculation on the part of Athenaeus. This is not probable though, and we are more likely to have the reason as stated by Polybius himself. This too could be pure speculation on the part of Polybius.

Some modern historians have suggested that the Games were held to celebrate the victories accomplished by Antiochus during his two invasions of Egypt carried out during the years 170-168. These invasions had been launched in response to Ptolemaic attempts to detach Koile-Syria from the Seleucid state. Although militarily successful, Antiochus had been forced to withdraw from Egypt by an ultimatum of the Roman Senate delivered to him at Eleusis near Alexandria by Popilius Laenas. As Bar-Kochva noted with respect to the first invasion of Egypt 'the rapid conquest of Egypt and the breakthrough westward from Pelusium .. had not been accomplished by any Hellenistic army since the days of Perdiccas'.[2] On his return to Antioch following the first invasion Antiochus had struck coins bearing the title of *Nikephoros*.[3] This epithet was retained in subsequent issues. As Mørkholm has pointed out:[4]

> 'From the Syrian point of view the results of the Sixth Syrian War might be regarded as not altogether unsatisfactory. The only battle during the

---

[1] Thus M. Rostovtzeff, *The Social & Economic History of the Hellenistic World*^{2} (Oxford 1953) II p. 699 describes it as Antiochus' 'retort to the Roman triumphs over the humiliated Hellenistic world'.

[2] Bar-Kochva, *Judas Maccabaeus* p. 112.

[3] Mørkholm, *Antiochus* p. 87.

[4] Mørkholm, *Antiochus* p. 97; that the Parade celebrated the victory in Egypt is also the view of Erich S. Gruen, *The Hellenistic World and the Coming of Rome* (Berkeley and Los Angeles 1984) II p. 660, who also holds that the military parade was also intended to show the increased strength of the Empire.

war had been a decisive Syrian victory. Egypt had been invaded twice and the Syrian army had returned intact and loaded with booty after the second expedition. More important was the fact that the question of Coele-Syria was now firmly settled in favour of Syria.'

Other historians have argued that in view of subsequent events it seems most probable that the festival was a prelude to Antiochus' Eastern Campaigns, designed to re-establish the cohesion of the Empire, rather than a retrospective celebration of the qualified victory in Egypt.[5]

In fact it would probably be a mistake to attribute to the Parade any one specific purpose to the exclusion of all others. Celebrations such as these, both in ancient and early modern times alike, although they could be occasioned by a particular event, such as a marriage or a victory, were used as an occasion to display whatever aspects of kingship the king and the dynasty wished to display. Thus it perhaps seems safest to follow the position of Bunge which is that the Parade was designed to celebrate all three events in the reign, both past and future ones.[6] The occasion of the Parade does not matter, what is of central importance is that the Parade was used to display the might, and especially the military might, of the Seleucid Kingdom.

With respect to the Parade Diodorus (31. 16. 1), presumably repeating Polybius, remarks that Antiochus adopted a policy opposite to that of the other kings. He continues.[7]

> ἐκεῖνοι γὰρ αὔξοντες τὴν ἰδίαν βασιλείαν καὶ δυνάμεσι καὶ χρημάτων πλήθεσι, καθὸ οἷοί τ' ἦσαν ἐπεκρύπτοντο τὴν αἵρεσιν διὰ τὴν Ῥωμαίων ὑπεροχήν· οὗτος δὲ τὴν ἐναντίαν λαβὼν διάθεσιν συνήγαγεν σχεδὸν ἀπὸ πάσης τῆς οἰκουμένης τοὺς ἐπιφανεστάτους ἄνδρας εἰς τὴν πανήγυριν, καὶ πάντα τὰ τοῦ βασιλείου μέρη διαφερόντως ἐκόσμησεν, εἰς ἕνα δὲ τόπον ἀθροίσας καὶ καθάπερ ἐπὶ σκηνὴν ἀναβιβάσας τὴν βασιλείαν ἅπασαν ἐποίησε μηδὲν ἀγνοεῖν τῶν περὶ αὐτόν.

> They, while strengthening their kingdoms both in arms and wealth, as far as possible tried to conceal their intentions because of the superiority of Rome. He, however, taking the opposite approach, brought together at his festival the most distinguished men from virtually the whole world, adorned all parts of the capital in magnificent fashion, and having as-

[5] Bar-Kochva, *Judas Maccabaeus* p. 470.

[6] J.G. Bunge, 'Die Feiern Antiochos' IV. Epiphanes in Daphnae im Herbst 166 v. Chr.' *Chiron* 6 (1976) pp. 53-71, esp. 70-71.

[7] Text and translation after Francis R. Walton in the Loeb series.

sembled in one spot, and, as it were, put upon the stage his entire kingdom, left them ignorant of nothing that concerned him.

The word translated as 'capital' here is an emendation by Salmasius of the manuscript βασιλέως to βασιλείου. This could be translated as either 'palace' or 'capital'. Given the context, Polybius presumably means 'capital' and implies that the festivities which followed the parade spread out into the city of Antioch rather than being confined to Daphne.

The date of the Parade has similarly been disputed. Athenaeus (10. 439 B) states that the description of the Parade comes from book thirty-one of Polybius' *Histories*. All other fragments from this book seem to be dateable to Ol. 154 (= 164-160 BC),[8] consequently most authorities consider the information supplied by Athenaeus to be incorrect.[9] Athenaeus could be correct only if the description of the Parade was given as a retrospective summary of the eccentricity and extravagances of Antiochus' reign, following a description of his death late in 164 BC. This seems extremely unlikely. Even if the purpose of the Parade, namely that it was held to outdo the triumphal parade and games of Aemilius Paulus in 167, is incorrectly given by Diodorus, this is in itself evidence that the Daphne Games took place soon after 167. In fact it seems probable that embassies were sent out to all the states of the Hellenistic world in the same year of 167 to invite them to send delegates to attend the festival, and that the festival itself took place in 166.

The Games must have taken place before Antiochus' departure for the east from Antioch. According to *1 Macc*. 3. 37 this took place in the year 147 SE. If the author is using the Macedonian calendar this would date to between Dios (October) 166 BC and Hyperberetaios (September) 165 BC, if he is using the Babylonian calendar it would date to between Nisan (April) 165 and Adar (March) 164. In this case he seems to be using the Macedonian dating, but certainty is impossible.[10] Mørkholm suggested that this dating entailed that the King set out on his eastern expedition in either the autumn of 166 or the spring of 165. Given that a Roman embassy headed by Ti. Gracchus visited Antiochus shortly after the Parade (Polyb. 30. 27; 30. 30) it is unlikely that Antiochus departed for the East immediately after the Parade. Mørkholm places the departure of Antiochus for the East in spring 165,[11] while Walbank dates Antiochus' departure to May 165.[12] Therefore most authorities place the Daphne Parade in the summer of 166.[13] Since Polybius can include events of the following autumn in his Olympiad year, the Parade could have

---

[8] Walbank, *Commentary* III pp. 30, 34-7.

[9] See Walbank, *Commentary* III p. 33.

[10] John R.Bartlett, *The Cambridge Bible Commentary, New English Bible. The First and Second Books of the Maccabees* (Cambridge 1973) pp. 6-7.

[11] Mørkholm p. 98 n. 37, p. 166.

[12] Walbank, *Commentary* III p. 33.

[13] Eg. Walther Kolbe in *Hermes* 62 (1927) pp. 238-9.

taken place either at the end of 167/6 or the beginning of 166/5.[14]

Bar-Kochva puts the Parade in the year 165. This seems most unlikely. Walbank has pointed out that this would entail a dislocation in the Polybian fragments. It would also necessitate the Parade taking place early enough in the year to allow Antiochus to depart for the east in May.[15]

Consequently it seems safest to accept the majority opinion that the Parade took place in 166. It seems probable, though not certain, that the Parade and subsequent festivities were simply an expansion of an annual festival in honour of Apollo which took place at Daphne. A more precise dating can only be achieved by assembling any evidence as to the season when the festival was normally celebrated. Downey has remarked that we do not know whether the Games held at Daphne were periodic or occasional.[16] By this he presumably does not mean to suggest that the festival might have been occasional, as this would be a most unusual state of affairs, but rather that the Games accompanying the festival may have been occasional. The festival itself would have been periodic: annual, biennial or penteric. *OGIS* 248, which probably records an Athenian decree crowning members of the Attalid royal house for the help which they had given to Antiochus IV in regaining his throne, states (vs. 51-3) that the award of the crowns is to be proclaimed in the contests celebrated in Pergamon and in those celebrated by King Antiochus at Daphne.[17] This surely suggests that the contests at Daphne were regularly celebrated.

Livy tells us that the festival and ritual games (sollemne ludorum) were also celebrated at Daphne in 195,[18] some 29 years earlier. For this reason the hypotheses that the festival was biennial or penteric are not sustainable, and it seems an inescapable conclusion that the festival was celebrated annually.[19] In 195 Hannibal met Antiochus son of Antiochus III at Daphne, where he was holding the ritual games.[20] Hannibal had left Carthage, or rather sailed from the island of Cercina 'media aestas forte erat' (Livy 33. 48. 5), which Holleaux interprets as meaning in

[14] Susan Sherwin-White and Amélie Kuhrt, *From Samarkhand to Sardis. A new approach to the Seleucid empire* (London 1993) p. 220 place the parade in the year 166/5.

[15] Bar-Kochva, *Judas Maccabaeus* pp. 468-73; Walbank *Scripta Classica Israelica* 15 (1996) p. 125 n. 39.

[16] Glanville Downey, *A History of Antioch in Syria from Seleucus to the Arab Conquest* (Princeton 1961) p. 94 n. 34.

[17] Cf. M.M. Austen, *The Hellenistic world from Alexander to the Roman conquest* (Cambridge 1981) no. 162.

[18] Livy places these games in 195, but Appian (*Syr.* 4) and Nepos (23 (*Hannibal*). 7) in 196. John Briscoe, *A Commentary on Livy Books XXXI-XXXIII* (Oxford 1973) p. 335 demonstrates that a date of 195 is by far more likely. Even if the date was 196, a penteric festival is still ruled out.

[19] Thus Mørkholm p. 98 n. 37 'It seems certain that the festival at Daphne was celebrated at fixed intervals, presumably every year'. I do not understand the statement of Briscoe *op. cit.* p. 335 that 'It is highly unlikely that such games were annual'.

[20] Livy 33. 49. 6. The Antiochus in question would have been Antiochus 'the prince', the eldest

July.[21] Livy (33. 49. 5) tells us that Hannibal arrived at Tyre 'prospero cursu', that is 'after a favourable journey', which one can presume to mean a swift and incident-free one. 'After a brief stay he sailed to Antioch' and met the prince at Daphne. According to Holleaux the journey would have taken three or four months, but this seems to be a considerable overestimate. If Hannibal's ship had indeed enjoyed favourable winds there is no reason to think that the whole journey would have taken more than a few weeks. Hannibal would therefore have been in Daphne in August, and so the Parade of 195 will have taken place in that month.

The Emperor Julian, writing in the fourth century AD, when speaking of Apollo at Daphne, tells us that 'In the tenth month, according to your reckoning - Loos I think you call it - there is a festival (ἑορτή) founded by your forefathers in honour of this god, and it was your duty to be zealous in visiting Daphne'. While it is true that the nature of the festival may have changed over time, there does not seem to be any reason to doubt that the festival was from its very beginnings traditionally celebrated in the month of Loos, which corresponds to August.[22] The festival of 166 would have taken place in August too.

A piece of evidence already mentioned suggests that the festival at Daphne was an annual rather than a penteric one. *OGIS* 248, an inscription found in Pergamon, but which Holleaux demonstrated is probably a copy of a decree passed by the city of Athens,[23] praises Eumenes II of Pergamon for the help he gave to Antiochus IV in re-gaining the throne and awards him with a gold crown. The award of the crown is to be proclaimed at the festivals celebrated in Athens and Pergamon and also (line 52) 'in those which King Antiochus celebrates at Daphne'. The decree would have been passed some time shortly after Antiochus' accession to the throne. We now know from a Babylonian document that Seleucus IV died on 3 September 175,[24] thus Antiochus' accession to the throne took place some time after this date, and thus too late to celebrate the festival at Daphne in 175. The Athenian decree would have been moved not too long after the accession, and thus

---

son of Antiochus the Great (as Briscoe, *op. cit.* p. 341) not the future Epiphanes (as Downey, *op. cit.* p. 94).

[21] Maurice Holleaux, *Études d'Épigraphie et d'Histoire Grecques. Tome V* (Paris 1957) p. 180-1.

[22] Julian, *Misopogon* 361 D; cf. E. S. Bouchier, *A Short History of Antioch 300 B.C. - A.D. 1268* (Oxford 1921) p. 49.

[23] See M. Holleaux, 'Un prétendu décret d'Antioche sur l'Oronte' *RÉG* 13 (1900) pp. 258-280, reprinted with additions by L. Robert in M. Holleaux, *Études d'Épigraphie et d'Histoire Grecques. Tome II* (Paris 1938) pp. 127-47. English translation of this inscription in M. M. Austin, *The Hellenistic world from Alexander to the Roman Conquest* (Cambridge 1981) no. 162.

[24] Sachs and Wiseman, *Iraq* 16 (1954) p. 208. See the treatment of this document by Mørkholm, *Antiochus* p. 43-3, who gives slightly different dates from the original publication on the basis of R. Parker and W. Dubberstein, *Babylonian Chronology 626 B.C. - A.D. 75 (= Brown University Studies 19*, Providence 1956) p. 23. The difference in dates is not significent for the current argument that Antiochus could not have been king in August 175.

Figure 33. *OGIS* 235 (photograph after Ulrich Köhler, 'Zwei Inschriften aus der Zeit Antiochus IV. Epiphanes' *Sitzungsbericht der Akademie der Wissenschaften Berlin* 51 (1900) pp. 1100-8.

very late in 175 or more probably early in 174. It is natural to assume that the decree anticipated the announcement of the award of a crown to Eumenes at the forthcoming celebration of the festival at Daphne in August 174. If the festival was an annual or biennial one then it would indeed have been celebrated in 174, but if it were a penteric one then the next festival would only have been celebrated in 172. This creates unacceptably long delays either between the accession of Antiochus and the decision of the Athenians to reward Eumenes for his help, or between the moving of the decree and the announcement of the reward of a crown at the next games at Daphne.

Bunge has, however, argued that the contests at Daphne took place in October 166, basing his argument on his interpretation of *OGIS* 253. This is an inscription carved on two contiguous fragments of white limestone, with lettering characteristic of the second century. Discovered at the end of the last century, it found its way to Berlin from Baghdad through the Antiquities market. The Baghdad dealer supplied the information that it had been found between Jumjuma (near Babylon) and Birs-Nimrud (Borsippa). Consequently, though the point is impossible to prove, it is generally assumed that the stone came from the city of Babylon.[25] Though lost

[25] Doubts on the recorded provenance have most recently been voiced by S. M. Sherwin-White, 'A Greek Ostrakon from Babylon of the Early Third Century BC' *ZPE* 47 (1982) pp. 51-70, esp. p. 65,

in the Second World War, a photograph of the inscription has been preserved which allows continued study of this important text. It has been known for a long time that the original restoration of Dittenberger in *OGIS* cannot stand,[26] and most recent work on the inscription uses the restoration of Zambelli, which is reproduced below.[27] The translation is based on that of Burstein.[28]

Βασιλεύοντος 'Αντιόχου θ[εοῦ 'Επιφανοῦς]
σωτῆρος τῆς 'Ασίας καὶ κτίσ[του καὶ εὐεργέτου]
τῆς πόλεως, ἔτους ςμ´ καὶ [ρ´ ---- θυσίαι? καὶ]
ἀγῶνι χαριστηρίοις ὑπο [------------------]
ἀπιόντος 'Υπερβερεταίου [βασιλεῖ 'Αντιόχωι]
θεῶι 'Επιφαν[εῖ ...] αν [--------------------]
Φίλιππος Δια[--------------------------]
ἐν τῶι δμ´ [καὶ ρ´ ἔτε--------------------]

In the reign of Antiochus, g[od manifest]
saviour of Asia and foun[der and benefactor]
of the city, year [1]46, [at the sacrifices(?) and]
contest of thanksgiving at [------------------]
in Hyperberetaios [------ to King Antiochus]
god manife[st---------------------------------]
Philippos son of Dia[-------------------------]
in [year 1]44 [---------------------------------]

The year 144 of the Seleucid era corresponds to the year 169/8 and the year 146 to 167/6. Thus we understand that the gist of the inscription concerns something that Philippos son of Dia[...] did in the year 169/8 and something that someone did in 167/6 at a contest of thanksgiving which seems to have taken place during the month of Hyperberetaios. Any assumption beyond this is uncertain.

Bunge proposed that line 4 be restored with the words ὑπὸ [Δάφνηι ἀγομένοις]. Thus he believed the thanksgiving contest referred to in that line were the games held at Daphne, which thus would have taken place during the last third of the month Hyperberetaios, that is in October 166.[29] This immediately creates one

---

repeated in Susan Sherwin-White & Amélie Kuhrt, *From Samarkhand to Sardis. A new approach to the Seleucid empire* (London 1993) p. 157.

[26] Doubts were first cast on the validity of Dittenberger's restoration by E. Bikerman, *Institutions des Séleucides* (Paris 1938) pp. 240-1 n. 9 and then by J. and L. Robert, *RÉG* 72 (1959) p. 149.

[27] Marcello Zambelli, 'L'Ascesa al Trono di Antioco IV' *Riv. Fil.* 88 (1960) pp. 363-389 esp. 378.

[28] Stanley Mayer Burstein, *The Hellenistic Age from the battle of Ipsos to the death of Kleopatra* = edd. E. Badian and Robert K. Sherk, *Translated Documents of Greece and Rome, Volume 3* (Cambridge 1985) no. 41.

[29] J. G. Bunge, 'Die Feiern Antiochos' IV. Epiphanes in Daphne' *Chiron* 6 (1976) pp. 53-71.

difficulty, for Polybius (30. 26) tells us that the games and associated activities lasted for thirty days, so it is likely that they spanned two months not one and thus two months would be mentioned in the inscription, for the restoration of which too little space would have existed in the inscription in its original complete state. Needless to say it also runs against the information which Julian gives us that the festival was celebrated in Loos. Furthermore, it seems most unlikely that the games regularly celebrated at Daphne would have been termed *Charisteria*, as, for whatever specific purpose they may have been staged in 166, they do not seem to have had the aim of celebrating a deliverance from anything. They rather have the nature of a national festival.[30]

It seems preferable to place more emphasis on the date of 169/8 given at the bottom of the inscription. Thus *OGIS* 253 should be interpreted as an inscription concerning *charisteria* which had been established in Babylon in the year 169/8, perhaps by Philippos son of Dia[...], and which were celebrated at the end of Hyperberetaios in the year 167/6, on which occasion the inscription was made.

Zambelli noted that the *charisteria* were celebrated at the same time of year as Antiochus' accession to the throne, and suggested that they were established to celebrate this event.[31] Mørkholm (p. 100) suggested that the title 'Saviour of Asia' given to Antiochus in the inscription, surely given in view of the king's victories in Egypt, ruled out this interpretation, and suggested that the *charisteria* were established to celebrate Antiochus' Egyptian victories, which saved the Seleucid state from its enemies.[32] The fact that Antiochus might be honoured with the epithet 'Saviour of Asia' in an inscription dating to after the Egyptian victory, in the view of the present author, in no way rules out the possibility that the *charisteria* were established to give thanks for Antiochus' deliverance of the same state from the hands of its (different) enemies in 175. Van der Spek accepted Mørkholm's interpretation and connected *OGIS* 253 with a Babylonian astronomical diary of the year 169/8 which tells us that 'King Antiochus entered victoriously into the cities of Egypt [...]. The *politai* [established] a *pompê* and a ritual (?) according to a Greek design [...]'. In the opinion of Van der Spek this text refers to the establishment of a festival by the Greek citizen population of Babylon to celebrate Antiochus' victory in the first successful campaign against Egypt in 169.[33] Once again, this is a

[30] Cf. Glanville Downey, *A History of Antioch in Syria from Seleucus to the Arab Conquest* (Princeton 1961) p. 98 n. 56.

[31] *Loc. cit.* pp. 378-80.

[32] Mørkholm, *Antiochus* p. 100. W. W. Tarn, *The Greeks in Bacrtia and India*[2] (Cambridge 1951) p. 154) also noted that Asia in the title 'Saviour of Asia' given to Antiochus in the inscription refers to the Seleucid Empire. His further suggestion (pp. 194-5) that the Festival at Daphne and the *charisteria* at Babylon were both celebrated to commemorate the victories of Eucratides over Demetrius of Bactria, and that the title of 'Saviour of Asia' was given to Antiochus because Eucratides was his general, has found minimal acceptance. See the comprehensive criticism of Tarn's views in A. K. Narain, *The Indo-Greeks* (Oxford 1957) pp. 54-7.

[33] R.J. van der Spek in edd. Amélie Kuhrt & Susan Sherwin-White, *Hellenism in the East* (London

possibility, perhaps even a probability, but it does not definitively rule out Zambelli's interpretation, as the two facts mentioned sequentially in the diary do not necessarily have to be connected.

The Philippos son of Dia[...] mentioned in the inscription may or may not be the 'minister for affairs' of Antiochus, whom Antiochus appointed regent on his deathbed, handing over to him his diadem, cloak and ring.[34] It may, however, be preferable to believe that he is an influential citizen of Babylon. The word Asia in the title 'Saviour of Asia' which is given to Antiochus in this inscription refers to the Seleucid Empire, and his role in the victory over the Egyptian invaders in 169.[35]

OGIS 253 should be rejected as a document referring to the contests held at Daphne, and a date of August 166 should be preferred for the festivities at Daphne.

1987) pp. 67-8. Contra M. J. Geller, 'New Information on Antiochus IV from Babylonian Astronomical Diaries' *Bulletin of the School of Oriental and African Studies* 54 (1991) pp. 1-4, who identifies the festival mentioned in the tablet with the Daphne Parade. However the fact that it is the (Greek) citizens of Babylon who celebrate the festival surely precludes the possibility that it it the Daphne Parade which is being mentioned in the tablet.

[34] The possibility was first noted by Dittenberger in *OGIS* ad loc. and later by Tarn, *op. cit.* p. 195. For doubts cast upon the suggestion see in the first place Walther Kolbe in *Hermes* 62 (1927) pp. 237-8.

[35] Tarn, *op. cit.* p. 153-4.

Appendix E

Polybius on Antiochus IV

Polybius is the paramount source for the history of the period, and his negative judgements have exercised an undue influence over later historians, ancient and modern alike. Most modern historians accept Polybius' account and opinion of the king at face value, and condemn him as mentally unstable. Even Mørkholm, the first modern historian to attempt a scientific enquiry into Antiochus' reign, pointed out how the opinions which modern historians have expressed about the nature and policy of the king are so strikingly divergent: Antiochus Epiphanes has remained an 'enigma' to modern historians.[1]

The partiality of Polybius has been observed by many modern historians as early as the last century.[2] Polybius was particularly hostile to the Antigonid monarchs Philip V and Perseus and to the Seleucid monarchs Antiochus III and IV. It is no surprise that as an influential citizen of the Achaean League Polybius would have been hostile to both these royal houses, as the state policy of the Achaean League was generally aligned to form alliances with the Ptolemies and against the Antigonids and Seleucids.[3] This in itself does not adequately explain the depth of Polybius' hostility. An interesting recent work of Walbank has helped to put his feelings into a rational framework.

Polybius wrote his Histories in order to investigate 'how in less than 53 years the Romans became masters of the inhabited world'. Like a multitude of modern historians in this century and the last, he sought some mechanism to explain the fall of one Empire and the rise of another. For Polybius it was the divine force of Fortune (*Tychê*) which was the mechanism which brought about the rise of Rome

---

[1] Mørkholm, *Antiochus* pp. 13-17; cf. Edwyn Robert Bevan, *The House of Seleucus* II (London 1902) pp. 131-2, Abel, *Revue Biblique* 50 (1941) p. 251.

[2] See e.g. J. P. Mahaffy, *A History of Egypt under the Ptolemaic Dynasty* (London 1898) p. 144-5 'Polybius is no more infallible in his judgement or free from prejudice than Thucydides, though both have that seriousness of style which averts suspicion'. According to Reuter (*Beiträge* p. 19) it was Wilamowitz who in 1908 first noted that Polybius' depiction of the history of the Seleucid state is inadequate and his judgement of the personalities involved unfair.

[3] Mørkholm, *Antiochus* p. 184; F. W. Walbank, *The Hellenistic World* (London 1981) p. 102. See Reuter, *Beiträge* pp. 19-20 on Polybius' local patriotism, the centrality of the Achaean League in his thinking, and his anti-monarchist tendencies. It is interesting to note how violently, when dealing with the Seleucid royal house, Polybius breaks his own rule (1. 14. 5) that the historian should free himself from attachment to friends and country (see Walbank, *Commentary* I p. 66 *ad. loc.*).

to supremacy. *Tychê* could either operate as an arbitrary force in human affairs, as, for example, in the case of natural disasters, or as a force exacting retribution. For Polybius *Tychê* is always a negative force, a reflection of the difficult world humans had to face.[4] Undoubtedly Polybius' personal experiences of the vicissitudes of fate as an exile had shaped his outlook. As far as Rome's victories over the Antigonid and Seleucid Kingdoms are concerned, Polybius (15. 20. 1-4) believed the cause to be the punishment meted out by *Tychê* for the monstrous crime committed by Philip V and Antiochus III in agreeing to dismember the inheritance of the infant Ptolemy V of Egypt. It is uncertain whether any formal agreement of this nature existed.[5] Some secret pact probably did exist, but this is in itself unimportant. All that matters is that Polybius believed it had. Not only were Philip and Antiochus to be punished, but the sins of the fathers were to be visited on the sons (15. 20. 5, 8).[6]

> οὐ μὴν ἀλλὰ τίς οὐκ ἂν εἰκότως τῇ τύχῃ μεμψάμενος ἐπὶ τῶν ἀνθρωπείων πραγμάτων ἐν τούτοις ἀντικαταλλαγείῃ, διότι ἐκείνοις μὲν ἐπέθηκε μετὰ ταῦτα τὴν ἁρμόζουσαν δίκην, τοῖς δ' ἐπιγενομένοις ἐξέθηκε κάλλιστον ὑπόδειγμα πρὸς <ἐπ>ανόρθωσιν τὸν τῶν προειπημένων βασιλέων παραδειγματισμόν

> But at the same time who among those who reasonably find fault with *Tychê* for her conduct of affairs, will not be reconciled to her when he learns how she afterwards made them pay the due penalty, and how she exhibited to their successors as a warning for their edification the exemplary chastisement she inflicted on these princes?

> τὸ τελευταῖον ἐν πάνυ βραχεῖ χρόνῳ τὴν μὲν Πτολεμαίου βασιλείαν ἡ τύχη διώρθωσε, τὰς δὲ τούτων δυναστείας καὶ τοὺς διαδόχους τοὺς μὲν ἄρδην ἀναστάτους ἐποίησε καὶ πανωλέθρους, τοὺς δὲ μικροῦ δεῖν τοῖς αὐτοῖς περιέβαλε συμπτώμασι ...

> And, finally, in a very short time *Tychê* re-established the kingdom of Ptolemy, while as for their dynasties and successors she in one case brought utter destruction upon them and in the other calamities very nearly as grave.

---

[4] Arthur M. Eckstein, *Moral Vision in The Histories of Polybius* (Berkeley 1995) p. 264.

[5] Cf. Erich S. Gruen, *The Hellenistic World and the Coming of Rome* (Berkeley and Los Angeles 1984) II pp. 387-8.

[6] Text and translation of these two passages as in the Loeb edition of W.R. Paton.

The concept of the retribution of *Tychê* as a mechanism of historical causality was not, however, designed by an impartial mind. Polybius had close connections with the Ptolemaic Court, and both he and his father Lykortas were politically active within the Achaean League in support of the policy advocated by Philopoemen of neutrality to Rome and close ties with the Ptolemaic and Attalid kingdoms. The ties between Lykortas and Polybius on the one hand and Philopoemen on the other were extremely close. When Philopoemen was killed in Messenia Lykortas was voted general and invaded Messene in order to recover the body. Philopoemen's body was cremated and his ashes were put in an urn decorated with wreaths and fillets. Polybius himself carried the urn back to Megalopolis at the head of the army (Plut., *Vit. Philop.* 21. 3). These events took place in 182.[7] The precise date of Polybius' birth is unknown, the earliest estimate puts it as early as 210/11, but he is more likely to have been born around 200[8] Polybius records an undated conversation between Philopoemen and Archon, the *stratêgos* of the League, at which he himself was present. The *stratêgia* of Archon can probably be placed in 190/89,[9] so Polybius would have been very young. In fact it is possible that Polybius and Philopoemen may have been related, which would explain the closeness of their relationship. Polybius' brother Thearidas called his son Philopoemen, which has led to the suggestion that Lykortas married Philopoemen's daughter.[10] It has been noted that if this were true Polybius would hardly have failed to mention it himself, and Plutarch would surely have mentioned it in his *Life of Philopoemen*.[11] If the two were related, their relationship would not have been that close.

The year 185 saw delegations from both Ptolemy and Seleucus arrive at the assembly of the Achaean League.[12] The Achaean League was currently the most powerful state on the Greek mainland and a valuable source of mercenary manpower. Recent relations between the Seleucid Kingdom and the League had not been good. In 190 a force of 1,000 Achaean infantry and 100 cavalry had been sent to help King Eumenes in the war against Antiochus the Great. They had fought against Seleucus himself at Pergamon, and had finally fought at Magnesia.[13]

On the other hand recent relations with Egypt had been good, and many of Egypt's mercenaries were recruited from within the territory of the Achaean League.

---

[7] R. M. Errington, *Philopoemen* (Oxford 1969) p. 245.

[8] Walbank, *Commentary* I p. 1 n.1.

[9] R. M. Errington, *Philopoemen* (Oxford 1969) p. 255.

[10] Stähelin, 'Lykortas' *PW* xiii, 2 (1927) 2386-9.

[11] Ziegler, 'Polybios (1)' *PW* xxi, 2 (1952) 1445; Walbank, *Commentary* I p. 228.

[12] R. M. Errington, *Philopoemen* (Oxford 1969) pp. 255-262, following Büttner-Wobst, argued that the embassy of Seleucus should be placed at the very beginning of his reign at the end of 187, which date has been accepted by Erich S. Gruen, *The Hellenistic World and the Coming of Rome* (Berkeley, Los Angeles, London 1984) II p. 645 n. 163. Against this chronology, however, see the arguments of Walbank, *Commentary* III pp. 9-10 supporting the chronology of Aymard.

[13] Polyb. 21. 3b; Livy 37. 20-21; 37. 39. 9

In 222 the mercenary garrison of Alexandria included nearly 3,000 Peloponnesians: Achaean mercenaries seem to have constituted the majority of the army, and a force of mercenaries raised in Greece is mentioned as having joined Ptolemy Epiphanes at Naucratis following the final suppression of the native rebellion.[14] Lykortas had been sent as ambassador to Egypt in 187/6, together with Theodoridas and Rositeles of Sikyon, in order to negotiate the renewal of the current alliance with Egypt,[15] and he returned with a Ptolemaic delegation bearing a gift of 6,000 bronze peltast panoplies and 200 talents weight of coined bronze to the League, requesting that the alliance sworn between Ptolemy and Lykortas be ratified.

The current *stratêgos* of the League, Aristainos, a political opponent of Philopoemen and his group, persuaded the assembly not to ratify the alliance, on the grounds that a number of alliances had existed with Ptolemy in the past, and both Ptolemy and the ambassadors had failed to state precisely which treaty was to be ratified. Undoubtedly Aristainos took this course of action for internal political reasons, in order to diminish confidence in Philopoemen.[16] The Seleucid delegation also wished to renew their alliance with the League and promised to give the Achaeans ten warships. The League voted to renew the alliance with Seleucus IV, but to refuse the ships. The refusal most probably reflects the policy of the League of maintaining the possibility of forming its foreign policy unimpaired by any moral commitments, which was the policy advocated by Aristainos and his party in opposition to that of Philopoemen.[17]

Subsequently, in 181/0, Ptolemy offered to give a squadron of pentekonteres, a gift worth not much less than ten talents, to the Achaean League. Having decided to accept this offer the League selected Lykortas and Polybius to thank Ptolemy for his previous gift of shields and coin and to receive the ships and organize their transport. As Polybius (24. 6. 4-5) reports:[18]

> κατέστησαν δὲ τὸν μὲν Λυκόρταν διὰ τὸ κατὰ τὸν καιρόν, καθ' ὃν ἐποιεῖτο τὴν ἀνανέωσιν τῆς συμμαχίας ὁ Πτολεμαῖος, στρατηγοῦντα τότε συνεργῆσαι φιλοτίμως αὐτῷ, τὸν δὲ Πολύβιον, νεώτερον ὄντα τῆς κατὰ τοὺς νόμους ἡλικίας, διὰ τὸ τήν τε συμμαχίαν αὐτοῦ τὸν πατέρα πρεσβεύσαντα πρὸς Πτολεμαῖον ἀνανεώσασθαι καὶ τὴν δωρεὰν τῶν ὅπλων καὶ τοῦ νομίσματος ἀγαγεῖν τοῖς 'Αρχαιοῖς.

[14] Polyb. 5. 36. 4; Launey I 137; Polyb. 22. 17. 6.

[15] Polyb. 22. 3. 6.

[16] Polyb. 22. 9, on the motives of Aristainos see R. M. Errington, *Philopoemen* (Oxford 1969) pp. 163-5.

[17] Polyb. 22. 7. 4; 22. 9. 13; R. M. Errington, *Philopoemen* (Oxford 1969) p. 162.

[18] Text and translation as in the Loeb edition of W. R. Paton.

They appointed Lykortas because, at the time when Ptolemy had renewed the alliance, he had been *stratêgos*, and had done his best to consult the king's interests, and Polybius, who had not attained the legal age for such a post, because his father had gone on an embassy to Ptolemy to renew the alliance, and to bring back the gift of arms and money.

This embassy never happened, however, due to the death of Ptolemy V Epiphanes at about this time. As the series of events clearly shows, Polybius had a very special connection with the Ptolemaic Court. Walbank has tentatively suggested that a relationship of *xenia*, or 'guest-friendship' may have existed between the family of Lykortas and the Ptolemies.[19] If so it is surprising that Polybius does not mention it at a relevant point in his text, but this does not necessarily rule out Walbank's suggestion.

Following the invasion of Egypt by Antiochus IV the Ptolemaic administration sent an embassy to the Achaean League in the winter of 169/8, begging them to send a military contingent of 1,000 foot and 200 horse, the whole force to be commanded by Lykortas and the cavalry by Polybius.[20] They also sent a message to Theodoridas of Sikyon, who had been ambassador to Egypt in 187/6 alongside Lykortas, begging him to raise a mercenary force of 1,000 men.[21] Of these three men, Lykortas, Polybius and Theodoridas, Polybius writes, 'The kings were particularly intimate with the men I have mentioned, owing to the circumstances narrated above'. Lykortas and Polybius both spoke in support of the delegation at the League's assembly in Corinth, and Polybius once again at a subsequent assembly at Sikyon.[22] On the second day Lykortas renewed the debate at Sikyon speaking for the Ptolemaic side. Polybius (29. 24. 16) says that Lykortas made a great impression.

καθ' ὅσον γὰρ οὐκ ἐξαριθμήσασθαι ῥᾴδιον ἦν τὰς τῶν ἐν 'Αλεξανδρείᾳ βασιλείων εὐεργεσίας, κατὰ τοσοῦτον ἁπλῶς οὐδὲν ἦν εὑρεῖν φιλάνθρωπον ἐκ τῆς 'Αντιόχου βασιλείας ἀπηντημένον εἰς πραγμάτων λόγον τοῖς 'Αχαιοῖς.

---

[19] F. W. Walbank, 'Supernatural Paraphernalia in Polybius' *Histories*' in ed. Ian Worthington, *Ventures into Greek History* (Oxford 1994) pp. 28-42, esp. p. 37.

[20] Polyb. 29. 23. 5.

[21] Polyb. 29. 23. 6; cf. Launey 1 p. 108-9 n. 6. It is not known whether this mercenary contingent was sent. The tradition of Achaean military help continued for a dedicatory inscription from Cyprus (*OGIS* 151; T.B. Mitford, *Opusc. Athen.* 1 (1953) p. 133 no. 4; *BSA* 56 (1961 p. 26 no. 69) dating to the middle decades of the second century was erected by the Achaeans and other Greeks campaigning on Cyprus (οἱ ἐν Κύπρω[ι] στρατευόμε[ν]οι 'Αχαιοὶ καὶ οἱ ἄλλοι Ἕλληνες) in favour of Seleucus son of Bithys a Rhodian (*PP* V *13790, VI *14951, VI 15078). It is highly unlikely that these troops are the remnants of the contingent raised by Theodoridas for Mitford has plausibly dated this inscription to 143 or early 142.

[22] Polyb. 29. 24. 1, 7.

For while it was not easy to enumerate the benefits conferred by the kings in Alexandria, there was not a single act of kindness of any practical value to be found which the Achaeans had met with from the dynasty of Antiochus.

Messengers now arrived from the Q. Marcius Philippus begging the Achaeans to follow the Roman policy of attempting to reconcile the two parties. Upon the arrival of this message the Ptolemaic envoys, realizing the Achaeans would send no military help, handed over a previously prepared letter from the two King Ptolemies, begging the Achaeans to send Lykortas and Polybius to help in the war.[23] They too were not sent.

Thus down to 167 when Polybius was interned in Italy as one of the thousand Achaean hostages sent to Rome, he was intimately involved in organizing support for the Ptolemies and action against Antiochus IV. Polybius' increasing lack of emotional detachment when discussing events of personal significance to his own life from book 29 onwards has already been noted.[24] This in itself would surely have been enough to guarantee Polybius' hostility to Antiochus in his *Histories*, but the contacts Polybius subsequently made in Rome, which constituted a major source of information available to the historian, seem to have been invariably hostile to Antiochus. Of the period between 167 and 150 which Polybius spent in Rome Walbank writes[25]

> 'In the exiles and in the numbers of countless embassies sent by their states to negotiate or plead with the Senate, the historian had an unparalleled source of information not only about current events but also about what had happened in previous decades, Since therefore most people who mattered at this time would turn up sooner or later in Rome, Polybius' exile gave him an opportunity which he would never have had in Achaea to learn not only about Rome but also about Greece and the Hellenistic kingdoms.'

It is safe to assume that Polybius got most of his information on affairs in the Seleucid state through the royal exile Demetrius and his circle in Rome. Demetrius the son of Seleucus IV was born in 186.[26] In 175 Seleucus was assassinated by his minister Heliodorus, who placed another son of Seleucus, named Antiochus, on the throne. Coin portraits of the young king show him to have been only 4-5 years old.[27]

---

[23] Polyb. 29. 25. 7.

[24] By Arthur M. Eckstein, *Moral Vision in The Histories of Polybius* (Berkeley 1995) p. 264.

[25] F.W. Walbank, *Polybius* (Berkeley and Los Angeles 1972) p. 10.

[26] Walbank, *Commentary* III p. 465.

[27] Otto Mørkholm, *Antiochus* p. 36.

Demetrius must, therefore, have been the eldest son of Seleucus,[28] and his legitimate successor. At this point, however, Antiochus IV seized the throne and adopted the infant Antiochus as his son, though subsequently murdering him.[29] Having pointed this out Mørkholm then went on to suggest that it would be reasonable to assume that Demetrius was hostile to Antiochus, whom he regarded as a usurper.[30] Paltiel argued that there is no direct evidence to suggest that any rivalry existed between Demetrius and Antiochus IV,[31] but nevertheless Mørkholm's assumption seems highly probable.

Demetrios had a circle of friends resident in Rome, some of whom were presumably exiles from Syria. These included his intimate friend and 'foster-brother' Apollonios son of Apollonios of Miletus,[32] and one Nikanor.[33] Demetrios periodically received visits from his adherents who would supply him with information on the current situation inside the Seleucid Kingdom. Upon the death of Antiochus IV at the end of 164 there was a political battle between Lysias and Philippos over which of the two would act as regent for his successor the infant Antiochus VI Eupator. Lysias won this battle, Philippos was killed, and it is reasonable to assume that many others fled the kingdom. One of these is the otherwise unknown Diodorus, the 'foster-father' of Demetrius, who arrived at Rome in 162 shortly before Demetrios' escape, who informs Demetrios on the current situation in Syria (Polyb. 31. 12. 3). There is no reason to think that Diodorus would have been hostile to Antiochus IV, as opposed to his obvious hostility to Lysias. Others in the circle around Demetrius who arrived before 164, however, may well have been hostile to Antiochus Epiphanes.

Thus, once again, it seems a reasonable assumption that Polybius received his information on what was happening in Syria from the circle around Demetrius.[34] Polybius was on extremely intimate terms with Demetrius. Indeed it seems to have been at the instigation of Polybius himself that Demetrius decided to escape from his captivity in Rome to take the throne from Antiochus' son and successor, Antiochus Eupator.[35] Having decided on escape, Demetrios begged for Polybius' help in organizing his escape. Polybius (31. 12. 8) describes his subsequent actions. At that

---

[28] Mørkholm, *op. cit.* p. 45 n. 31.

[29] Mørkholm, *op. cit.* pp. 41-47

[30] Mørkholm, *op. cit.* p. 184.

[31] Eliezer Paltiel, 'Antiochus IV and Demetrius I of Syria' *Antichthon* 13 (1979) pp. 42-47.

[32] Polyb. 31. 11. 7.

[33] Polyb. 31. 14. 4. Nothing further is known of this Nikanor. He is no evidence to connect him with Nikanor son of Patrokles, governor of Palestine under Antiochus IV (Th. Lenschau, 'Nikanor (16a)' *PW Supplbd.* vii (1940) 568-9; *2 Macc.* 8. 9; *1 Macc.* 3. 38 - 4. 25; Joseph., *AJ* 12. 402-412; *2 Macc.* 8. 8-29).

[34] Paul Pédech, *La méthode historique de Polybe* (Paris 1964) p. 271 considered it probable that Diodorus and Apollonius were oral sources; Mørkholm, *Antiochus* p. 184.

[35] F. W. Walbank, *Polybius* (Berkeley and Los Angeles 1972) p. 9.

time Ptolemy Euergetes was in Rome petitioning the Senate that Cyprus be assigned to him, the Ptolemaic Empire having been partitioned between him and his elder brother Ptolemy Philometor.[36]

> συνέβαινε δὲ κατὰ τὸν καιρὸν τοῦτον Μένυλλον μὲν τὸν Ἀλαβανδέα παρεῖναι πρεσβεύοντα παρὰ τοῦ πρεσβυτέρου βασιλέως Πτολεμαίου χάριν τοῦ συγκαταστῆναι καὶ δικαιολογηθῆναι πρὸς τὸν νεώτερον Πτολεμαῖον· ἐτύγχανε δὲ προϋπάρχουσα τῷ Πολυβίῳ πρὸς τὸν Μένυλλον ἰσχυρὰ συνήθεια καὶ πίστις.
>
> At that time it happened that there was a certain Menyllus of Alabanda present, on an embassy from the elder Ptolemy, with the object of confronting and answering the younger Ptolemy. Polybius had long been intimate with this Menyllus, and had great confidence in him.

It was Menyllus who organized the ship in which Demetrios escaped. Menyllus had probably accompanied the older Ptolemy Philometor when the king made a personal visit to Rome in 164 to appeal for help when he had been expelled from Alexandria by his brother Euergetes. The Seleucid prince Demetrios met Philometor as he approached Rome on foot, greeting him and offering him royal regalia. Philometor declined this gift and approached Rome in the dress of a commoner accompanied by only one eunuch and three slaves. He took up lodgings with the Alexandrian landscape-painter Demetrios, whom the king had often entertained in Alexandria. All these steps were taken to arouse the sympathy of the Senate. It is unthinkable that Polybius would not have met the King during his visit of 164.[37] Polybius probably accompanied Demetrios when he met Philometor on the road to Rome. It was probably on this occasion that Polybius met Menyllus of Alabanda.

We have seen that Polybius was in any case already pre-disposed to take a very negative view of Antiochus IV Epiphanes due to his political activities prior to his exile in 167. As to the details of Antiochus' activities within the Seleucid Kingdom, such as his administrative and military reforms, Polybius was reliant on the information retailed by Demetrius, his circle, and his visitors. As to the details of what had transpired between Philometor and Antiochus during the Seleucid double intervention in Egypt in 170-168, Polybius was reliant on what Philometor and Menyllus of Alabanda told him.[38]

---

[36] Text and translation from the Loeb edition of W. R. Paton.

[37] The events are described in Diod. 31. 18. On this passage see Edwyn Bevan, *A History of Egypt under the Ptolemaic Dynasty* (London 1927) p. 291 n. 2 'Probably he was drawing from Polybius, who was in Rome at the time, and no doubt knew Philometor there'. See also Mørkholm, *Antiochus* p. 184 on the probability that Polybius met Philometor in Rome.

[38] Mørkholm, *Antiochus* p. 184 notes that Polybius was 'on intimate terms with' Menyllus of Alabanda and other Ptolemaic officials.

Thus we should expect Polybius to give a most unsympathetic account of Antiochus Epiphanes. If we examine Polybius' account of the death of Antiochus (31. 9) as an example, we can see this animus in operation. Polybius' description of the death of Antiochus IV has survived in the following excerpt (31. 9. 1-4)

> Ὅτι κατὰ τὴν Συρίαν Ἀντίοχος ὁ βασιλεὺς βουλόμενος εὐπορῆσαι χρημάτων προέθετο στρατεύειν ἐπὶ τὸ τῆς Ἀρτέμιδος ἱερὸν εἰς τὴν Ἐλυμαΐδα. παραγενόμενος δ' ἐπὶ τοὺς τόπους καὶ διαψευσθεὶς τῆς ἐλπίδος διὰ τὸ μὴ συγχωρεῖν τῇ παρανομίᾳ τοὺς βαρβάρους <τοὺς> οἰκοῦντας περὶ τὸν τόπον, ἀναχωρῶν ἐν Τάβαις τῆς Περσίδος ἐξέλιπε τὸν βίον, δαιμονήσας, ὥς ἔνιοί φασι, διὰ τὸ γενέσθαι τινὰς ἐπισημασίας τοῦ δαιμονίου κατὰ τὴν περὶ τὸ προειρημένον ἱερὸν παρανομίαν.

> In Syria, as King Antiochus wished to have plenty of money, he decided to make a campaign against the sanctuary of Artemis in Elymaïs. Arriving at the place he was cheated of his hope, for the barbarians living around the place would not consent to the transgression, and, retreating, he died at Tabai in Persis, possessed by a madness, as some say, on account of certain divine manifestations which happened during the transgression with regard to the aforementioned sanctuary.

There are a number of problems with this passage, most of which are not relevant to the present discussion. By using the word στρατεύειν Polybius implies, and the very late accounts of this incident, under his influence, state, that Antiochus attempted to take the temple treasures by force. Had Antiochus in fact attempted this at the head of his army, there would have been little that the barbarians living around the place could have done to prevent it, as was pointed out by Tarn.[39] Tarn further notes that whatever transgression it was to which the local inhabitants objected, it must have been 'a peaceful dealing and not an attack'. It is worth noting that Antiochus did not press his case when his proposal was refused. Although it is quite possible that the ultimate aim of Antiochus was to appropriate the temple treasures following a sacred marriage, the reason alleged in the excerpt of Polybius, that Antiochus wanted to enrich himself, remains no more than a disparaging allegation on the part of Polybius.

It is not the purpose of this work to enter into a discussion of the role of temple-robbing in Seleucid state policy; however it is worth noting the observation of Le Rider that temple-robbing by the later Seleucid monarchs was more a politico-military act than an indication of poverty.[40] Indeed the aim of much of this tem-

---

[39] W. W. Tarn, *The Greeks in Bactria and India2* (Cambridge 1951) pp. 463-8.

[40] Georges Le Rider, 'Les resources financières de Séleucos IV (187-175) et le paiement de l'indemnité

ple-robbing may have been to deny potential rebels the possible use of these financial resources in the future. By suppressing any discussion of the circumstances surrounding Antiochus' transaction with the temple, Polybius seeks to create the impression that Antiochus is nothing more than a temple-robber, one of the worst types of criminal known in the Greek world.

Polybius then has Antiochus retreat to Tabai where he dies 'as some say' struck down with madness due to some manifestations of divine displeasure when he was attempting this outrage on the sanctuary. Presumably Polybius' source for his lurid account of Antiochus' death was hostile religious propaganda in existence at the time, put into circulation by Antiochus' enemies. Polybius has chosen to include the unhistorical to colour his account of Antiochus' death.[41] In fact we do not know the true cause of Antiochus' death. It could be argued that Polybius sees the hand of *Tyche* in the death of Antiochus, and this would fit in with his causal theory of history. As in many of the cases where he sees the supernatural at work to punish the guilty, he deliberately uses the term ὡς ἔνιοί φασι ('as some say') to describe Antiochus' death, struck down by madness.[42] In the opinion of the present author it is primarily simple animosity to Antiochus which is at work in this passage, and not belief in the operation of *Tychê*. For this reason Polybius leaves what actually happened at the temple unclear.

It is not usual to find Polybius distorting his reportage to this extent with gossip and supernatural intervention, and perhaps even deliberately suppressing material. We have to be very careful when dealing with his treatment of Antiochus. Polybius had been a politician and soldier during the Seleucid invasion of Egypt, when he had taken an active part in events, supporting the Ptolemaic interest. Doubtless the speeches made by Polybius and others during these events had elaborated the Ptolemaic view that it was Antiochus who had broken faith and initiated the Sixth Syrian War. They would also have dwelt on Antiochus' alleged dubious character. Evidently Polybius had begun to believe his own propaganda. When writing his version of events in Rome he was furthermore dependent on hostile sources. For these reasons Polybius' reportage of the doings of Antiochus is exceptionally hostile and dismissive, and we should exercise due care when using them.

Nevertheless even Polybius felt himself forced to report a more favourable opinion of Antiochus.[43] Although criticizing his management of Antiochus' campaign near Pelusium, Polybius (28. 18. 1; cf. Diod. 30. 18. 1) reports that the king was 'active, capable of forming great designs, and worthy of the royal dignity' (καὶ πρακτικὸς καὶ μεγαλεπίβολος καὶ τοῦ τῆς βασιλείας προσχήματος ἄξ-

---

aux Romaines' in edd. Martin Price, Andrew Burnett and Roger Bland, *Essays in Honour of Robert Carson and Kenneth Jenkins* (London 1993) esp. p. 61.

[41] P. Pédech, *La méthode historique de Polybe* (Paris 1964) p. 395; Mørkholm, *Antiochus* p. 183.

[42] F. W. Walbank, *Polybius* (Berkeley and Los Angeles 1972) p. 65.

[43] As noted by Mørkholm, *Antiochus* pp. 183-4

ιος). Similarly in a passage devoted to describing the king's eccentricities, Polybius (26. 1. 10) tells us that he far surpassed his predecessors in the honours he paid to the gods. It is notable that Polybius fails to mention anything unusual in Antiochus' attitudes to the gods, which casts doubt on suggestions that Antiochus was a religious revolutionary. In a second passage Polybius (29. 24. 13), in recording a speech of his father Lykortas to the Achaean assembly in the winter of 169/8, has him admit that Antiochus IV, unlike his ancestors, had acted with conspicuous magnanimity (μεγαλοψυχία) towards the Greeks.[44] Livy (41. 20. 5), in a passage certainly derived from Polybius,[45] tells us that in two great respects Antiochus' soul was truly royal: in his benefactions to cities and in the honours paid to the gods. A long list of his benefactions follows with the statement that he promised to do many other things in many other places but due to the very short duration of his reign he did not finish them. Livy (41. 20. 10) then mentions that he also surpassed all his predecessors in the splendour of the shows of all sorts he organized.

This positive opinion is also reflected in Appian (*Syr.* 45), who uses sources independent of the Polybian tradition and more friendly to the Seleucid royal house.[46] Appian tells us that the Syrians themselves named him Epiphanes because when the rule was taken by strangers he proved himself to be a proper king. Appian continues by stating that Antiochus, having affirmed his friendship and alliance with King Eumenes he ruled Syria and the lands about it firmly. As far as the epithet 'Epiphanes' is concerned, Nock has noted

> "its convenient width of meaning, ranging from 'god showing himself' to 'distinguished,' suited well the vagueness characteristic of the language employed in this connection. Antiochus IV adopted it, and then defined its meaning more closely as ΘΕΟΣ ΕΠΙΦΑΝΗΣ."[47]

---

[44] On this passage see Reuter, *Beiträge* p. 17; the benefactions of Antiochus are listed in Mørkholm, *Antiochus* pp. 58-63; Erich S. Gruen, *The Hellenistic World and the Coming of Rome* (Berkeley, Los Angeles, London 1984) II p. 647.

[45] Walbank, *Commentary* III p. 287.

[46] Mørkholm, *Antiochus* pp. 19, 184-5; Kai Broderson, *Appians Abriss der Seleukidengeschicht (Syriake 45, 232-70, 369) Text und Kommentar etc.* (Munich 1989) p. 90 remarks on the problematic nature of the sources used by Appian. Although the *Histories* of Poseidonius of Apamea only begin where Polybius leaves off in 146, they could contain material favourable to Antiochus included as a digression during the course of describing the civil wars in Syria in the later second century between the descendants of Antiochus Epiphanes and those of Demetrius. Cf. the discussion in Paul Pédech, *La méthode historique de Polybe* (Paris 1964) pp. 270-1 of the sources on Syrian history which may have been available to Polybius.

[47] A.D. Nock, 'Notes on Ruler-Cult, I-IV' *JHS* 48 (1928) 21-43 (= ed. Z. Stewart, *Arthur Darby Nock. Essays on Religion and the Ancient World etc.* (Oxford 1972) I pp. 134-159) esp. pp. 40-41. Cf. Otto Mørkholm, *Studies in the Coinage of Antiochus IV of Syria* (Historisk-filosofiske Meddelelser udgivet af Det Kogelige Danske Videnskabernes Selskab, Bind 40 nr. 3, Copenhagen 1963) p. 71 f. 'Antiochus was presented to the Greeks in the west with an epithet of a convenient vagueness, which need not imply deification, while to his own subjects he was undoubtably the God Manifest'. See also

Thus Appian may be correct in suggesting that the title was originally adopted with reference to the legitimacy of Antiochus to rule.[48] One might doubt his assertion that the Syrians gave Antiochus the title, rather than it being adopted by the king himself. Appian (*Syr.* 46) then goes on to say that after Antiochus died the Syrians gave the name of Eupator to his son on account of his father's virtue. There can be no room for doubt in the case of Eupator. The latter was only an infant and would hardly have decided on the title himself. In fact *1 Maccabees* (6. 17) says that the regent Lysias gave Eupator the title himself. Whichever may be the case, the title attests to the good feelings of the Syrian population towards Epiphanes.[49]

Even later on the memory of Antiochus Epiphanes as a good king was exploited for the purposes of political propaganda over the next 20 years in the convulsive dynastic struggles which beset the Seleucid realm. Alexander Balas, who claimed to be a son of Antiochus, invented the title of Theopator for use on his coins,[50] and towards the end of his reign he issued commemorative coinage with the portrait and title of Antiochus Epiphanes. Similar commemorative coinage was issued upon the assumption of the Seleucid throne by his son Antiochus VI. All this 'leaves no doubt that his memory was still alive and cherished with affection by at least a substantial proportion of the Syrian population'.[51] The impression that Antiochus was popular among his own subjects is even given by the generally hostile *1 Macc.* 6. 11 which has Antiochus on his deathbed ask why he should be overwhelmed by a flood of trouble 'I who was kind and well-loved in the day of my power?'.[52]

How, therefore, are we to treat the statement of Polybius that because of his acts Antiochus came to be called 'Epimanes', or Insane, rather than 'Epiphanes' or Illustrious?[53] Polybius retails this information in the course of his discussion in book 26 of how Antiochus' eccentricities were viewed by the people of Antioch, implying that the Antiochenes themselves thought him mad and gave him the epithet, though not actually stating this as a fact. In consequence of this artfulness on the part of Polybius, a number of modern scholars have assumed that the epithet was

---

Walbank, *Commentary* III p. 285 for further discussion of the epithet.

[48] Cf Broderson, *op. cit.* p. 63 'Appians Erklärung betont also die Legitimität der Herrschaftsfolge'.

[49] Cf. Broderson, *op. cit.* p. 67.

[50] On this title see Georges Le Rider, *Suse sous les Séleucides et les Parthes* (= *Mémoires de la Mission archéologique en Iran* 38, Paris 1965) p. 316 n. 7.

[51] Mørkholm, *Antiochus* p. 185; see also Reuter, *Beträge* pp. 17-19; Arthur Houghton, *Coins of the Seleucid Empire from the Collection of Arthur Houghton* (New York 1983) nos. 209-213.

[52] Noted by Edwyn Robert Bevan, *The House of Seleucus* (London 1902) II p. 131 n. 2.

[53] Polyb. 26. 1a. 1, 1. 1 in Athen. 10. 439 a, cf. 5. 193 c-d, 5. 196 a, cf. Livy 41. 20. 4. Reuter (*Beiträge* pp. 26-7) noted that this piece of information did not come from Ptolemy Euergetes. He tried to save the reputation of Polybius by suggesting (*Beiträge* 27 n. 45) that Athenaeus should be emended to have Polybius report that Epiphanes was called Epimanes, rather than stating it himself.

[54] Most recently in Peter Green, *Alexander to Actium. The Hellenistic Age* (London 1990) p. 437 'Wits in Antioch'.

given him by the populace of Antioch.[54] In fact it is much more probable that Antiochus' political enemies, more specifically the coterie surrounding the future Demetrius I in Rome,[55] of which Polybius himself was a member, invented this nickname for Antiochus and did their best to make it stick even after his death. Hence its appearance in the text of Polybius.[56]

---

[55] Thus Abel, *Revue Biblique* 50 (1941) pp. 236-7.

[56] Paul Pédech, *La méthode historique de Polybe* (Paris 1964) p. 150 n. 270 noted that the adjective epimanes first appears in Polybius.

Figure 34. The Kampyr-Tepe terracotta plaque
(photograph after *Iran* 30 (1992) pl. 14 a).

Appendix F

The Kampyr-Tepe Terracotta

A further piece of archaeological evidence, though not Seleucid, should perhaps be considered at this point. Figure 34 shows a terracotta plaque from Kampyr-Tepe. It depicts a heavily-armed warrior striding to his left and standing under an arch. An interpretation and reconstruction of this figure has already been undertaken by Nikonorov.[1] On his head he wears a helmet. The shape of the helmet cannot be made out in precise detail, but it seems to have an extended skull ending in a reinforcing plate at the top of the skull surmounted by a ball. A fillet possibly falls behind the ball of the helmet, but this detail is unclear. There may be a nasal or other plate-like attachment at the front of the skull of the helmet, but the detail cannot fully be made out. He wears a cuirass which Nikonorov interprets as a plain cuirass without the musculature typical of the muscle-cuirass, however one should not necessarily expect the musculature to be shown in an object of such small size. Nikonorov interprets a shape seen behind the neck as a neck-protector. The arms seem to be bare. Around his neck he wears what appears to be the fold of a cloak pushed back over both shoulders, as Nikonorov has restored in his reconstruction. In his right hand he holds a sword with a diamond-shaped blade. The muscle-cuirass is straight at the bottom where it seems to end in a rolled edge, below which fall one short and one long row of *pteruges*. The legs and feet appear bare, but Nikonorov has restored greaves and sandals. In his left hand he holds a large oval *thureos* shield with a design which might be interpreted as a winged thunderbolt, which interpretation Nikonorov rejects, seeing in the device a lizard or perhaps a *varan*.

Pugatchenkova considered that the plaque depicts a Roman soldier and perhaps dates to the first half of the second century AD.[2] Nikonorov and Savchuk pointed out that this was unlikely for chronological reasons. It was found in a deposit from Tower 2 on the citadel, above which there were two layers, above which in turn was discovered a silver coin of the Graeco-Bactrian King Eucratides (c. 170-145 BC) in the plaster of a wall levelled in the Kushan period. 'Thus, the reign of Eucratides is likely to be the latest period for our terracottas'.[3] Nikonorov be-

[1] Valerii P. Nikonorov, *The Armies of Bactria 700 BC - 450 AD* (1997) Vol. 1 p. 38-9, Vol. 2 p. 10, pl. 3A, fig. 24a.

[2] G.A. Pugachenkova in *Obshchestvennye nauki v Uzbekistanie 1989*, no. 4 pp. 55-7.

[3] V.P. Nikonorov and S.A. Savchuk *Sovietskaya Arkheologia* 1992, p. 50 fig. 2; Valeri P. Nikonorov and Sergei A. Savchuk, 'New Data on Ancient Bactrian Body-Armour (in the light of Finds from

lieves the terracotta shows a heavy infantryman of the Graeco-Bactrian kingdom.

It is the purpose of this Appendix to tentatively advance the suggestion that this plaque may possibly depict a Roman-armed infantryman of the first half of the second century BC. If this suggestion is correct, then the 'spiked' helmet could be a local type based on the Montefortino type. It has the high, domed, skull of the Montefortino type, though the shape of the cheek-pieces can not be made out. The figure seems to wear a typically Greek muscle-cuirass with *pteruges*. This does not hinder our interpretation however, for representations of Italian infantry continue to show them wearing the muscle-cuirass down into the first century, and a large number of muscle-cuirasses have been excavated from Campanian graves of this period.[4] He carries a large oval *thureos* shield. In his right hand he carried a wide-bladed sword in the shape of an extended lozenge. Nikonorov believed the sword to be Greek in type not Roman. He could be wearing greaves or boots on his legs, but the detail cannot be made out sufficiently

As Nikonorov notes, the figure does not represent a Macedonian-style phalangite, since the *thureos* would not permit him to use the sarisa. Normally if Hellenistic infantry carried a *thureos* (i.e. were infantry of the type called *thureophoroi*) during the third and early second centuries BC they would act as medium infantry, without cuirass or greaves, but this warrior is decidedly 'heavy'. The fact that the warrior wears a mixture of Greek and 'Roman' equipment is not an obstacle to an interpretation of the figure as a Roman-armed infantryman, Senior officers in the Roman army wore muscle cuirasses with *pteruges* and greaves, and this figure could be a senior officer.

There are opportunities after the collapse of the Bactrian kingdom when Seleucid troops could have been operating in the area. For example after the death of Antiochus VII Sidetes and the collapse of his campaign to recover the eastern provinces the Parthian King Phraates II captured a sizeable part of his huge army. Justin (42. 1. 3) tells us that when Phraates led this captive army eastwards against the Sakai they turned traitor and accomplished the destruction of the Parthian army and the death of Phraates himself. The date of these events, 129 BC, is probably too late for the date of the terracotta however.

If the proposition advanced above is correct and this does represent a Roman-armed infantryman, then the probability must be that the warrior represented on the plaque must belong to the Graeco-Bactrian army, perhaps of the reign of Eucratides the Great (c. 170-145 BC), which is where Nikonorov placed him. It would imply that the Graeco-Bactrian army also attempted military reform along Roman lines. The tentative nature of this suggestion cannot be emphasized too strongly. One should resist the temptation of erecting a complex intellectual structure on the foundation of one terracotta.

Kampyr Tepe)' *Iran* 30 (1992) pp. 49-54 pl. 14 esp. p. 50 no. 1 pl. 14a.

[4] Nick Sekunda, Angus McBride, *Republican Roman Army 200-104 BC* (London 1996) pp. 8-9.

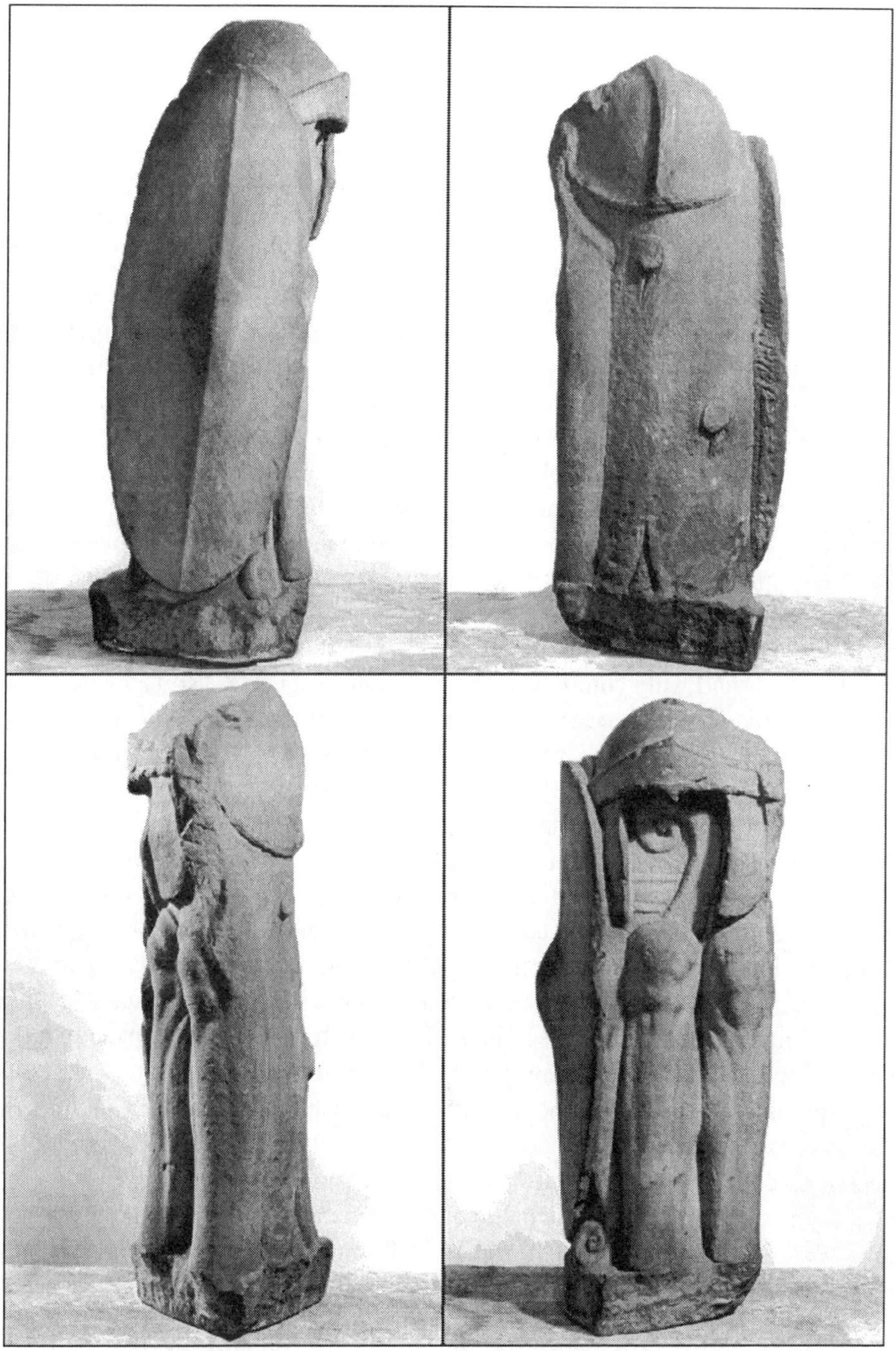

Figure 35. Marble trophy in the British Museum, Department of Greek and Roman Antiquities, inv. sculpture 2142 (photograph: Courtesy of the Trustees of the British Museum).

Appendix G

Late Hellenistic Armies

By the time of Mithridates the Great troops armed in the Roman manner were starting to make a more regular appearance in Hellenistic armies. During the First Mithridatic War the Pontic army fielded a mixed infantry force. In the campaign of 86 BC one part was equipped in the Macedonian manner and the other in the Roman. Frontinus (*Strat.* 2. 3. 17) tells us that at the battle of Orchomenos the Mithridatic general Archelaos placed scythe-bearing chariots in the front line, then after them the 'Macedonian phalanx' in the second line. In the third line he placed auxiliaries 'armed in the Roman manner' with Italian fugitives mixed in with them, then light infantry in the last line and cavalry on either flank. The *Chalkaspides* mentioned by Plutarch (*Vit. Sull.* 16. 7) earlier on in the campaign are the Macedonian phalanx element of the infantry. They fought with long *sarisai* and locked shields (18. 4) and were commanded by Taxiles (19. 2). The 'Romanized' formation will be the 15,000 freed slaves mentioned by Plutarch (*Vit. Sull.* 18. 5).

There is no archaeological evidence concerning the Mithridatic army during the Boeotian campaign of 86 BC which can be used with any certainty. A marble trophy (Figure 35) was presented to the British Museum by John Walker Esq. in 1802, supposedly having come from the Plain of Marathon. The trophy, which originally served as a support for a statue which stood at its side, is 84 cm high, and depicts a helmet with cheek-pieces, a bow, greaves and a shield all on a tree-trunk. Vanderpool thought the attribution to the Plain of Marathon might have been false and compared the shield and helmet with those shown on the Domitius Ahenobarbus monument in the Louvre (see Figure 18).[1] In other words he considered these weapons to have been of Roman type. A later treatment of this same monument suggested that it may commemorate the battle of Chaironeia in 86 BC,[2] in which case the helmet, the shield and possibly the greaves may represent components of the equipment of the 'Romanized' contingent of the Mithridatic army.

In the closing stages of the battle of Orchomenos elements of the Mithridatic army were forced into the marshes of Lake Copaias. Plutarch (*Vit. Sull.* 21. 4), a native of the neighbouring city of Chaironeia, tells us that even in his day many

---

[1] Eugene Vanderpool, 'The Marble Trophy from Marathon in the British Museum' *Hesp.* 36 (1967) pp. 108-111, pl. 31.

[2] John Camp, Michael Ierardi, Jeremy McInerney, Kathryn Morgan and Gretchen Umholz, 'A Trophy from the Battle of Chaironeia of 86 BC' *AJA* 96 (1992) pp. 443-455.

Figure 36. Helmet found east of Haliartos in National Museum Athens inv. X 15117
(photograph: National Museum, Athens).

barbarian bows, helmets, fragments of iron cuirasses and swords were found in the mud of the Lake. When Plutarch mentions fragments of iron cuirasses he may be talking about mail cuirasses of the Roman type. It would be tempting to associate weapons recovered from the area of the lake in modern times with this battle. Calligas, in addition to the Zagora Polydrossou helmet, mentions a further four helmets of the Montefortino type recovered in Greece, plus cheek-pieces of this type from Dodona, which have already been mentioned, and from Corcyra.[3] Two of these helmets come from the Lake Copaias region, and could conceivably come from the battle.

The example illustrated here (Figure 36) was given to the Museum in 1926 and was reported to have been found some 7-8 km. east of Haliartos. It was therefore found in the wrong place and it also seems to be too early in date. Down to the second century the bowls of the helmet had been well decorated and finished, as in this example, but there was a marked deterioration in the quality of manufacture of these helmets at the end of the second century. This was due to two reasons. Firstly the size of the armies put in the field by the Roman state increased, and so the number of helmets required went up. Furthermore previously it was the duty of the conscript to supply his own equipment, and so the helmets were private property bought by an individual. During the second century the state increasingly started to provide the Roman conscripts with their equipment, and helmets were increasingly purchased and manufactured on a mass scale.[4] A first century date for this helmet is precluded for this reason. Calligas is probably correct in dating both these helmets to the turn of the third and second centuries and in associating them with the campaigns of the Roman army in the area during the Macedonian Wars. It could be argued that the helmet is even earlier and might have been deposited during the Galatian incursions into Central Greece in 279 BC. This is ruled out by the fact that the helmet has a three-letter inscription in the Latin alphabet. The other two helmets listed by Calligas come from Locris and from an old collection in Thessaly, perhaps coming originally from a shipwreck.

Following his defeat Mithridates raised a new army. Plutarch (*Vit. Luc.* 7. 4) tells us that he had swords and shields of Roman type prepared and a force of 120,000 men trained in Roman tactics. According to Appian (*Mith.* 87, 108) this later Mithridatic army was divided into cohorts (*speirai*) of 600 men along Italian lines. The process of 'Romanization' spread to the Armenian army too. At the battle of Tigranocerta in 69 BC the 150,000 strong Armenian heavy infantry was drawn up 'some in cohorts and others in phalanxes' (Plut., *Vit. Luc.* 26. 6 - τῶν μὲν εἰς σπείρας, τῶν δ' εἰς φάλαγγας συντεταγμένων).

---

[3] P.G. Calligas, 'Roman Helmets in Greece' *AAA* 18 (1985) pp. 161-4.

[4] John Paddock, 'Some changes in the manufacture and supply of Roman helmets under the late Republic and early Empire' in ed. M. C. Bishop, *The Production and Distribution of Roman Military Equipment* (=British Archaeological Reports, International Series 275, Oxford 1985) pp.142-59, esp. 144-5.

By the time of the Civil Wars, it seems that all major remaining armies of the Hellenistic world had converted to Roman equipment and tactics. Thus the army that Juba fielded against Caesar was organized in legions and cohorts (Caes., *Bell. Afr.* 48, 55, 59). In Egypt considerable power lay in the hands of the Gabiniani, Gauls and Germans, whom the governor of Syria A. Gabinius had left in Egypt to maintain Ptolemy Auletes on the throne.[5] At the same time Dieotarus tetrarch of Galatia had organized two legions of Galatians equipped and trained along Roman lines.[6]

[5]Caes., *Bell. Civ.* 3. 4. 4

[6]Caes., *Bell. Alex.* 34. 4, 68; cf. Cic., *ad Att.* 6. 1. 14 where Deiotarus has 30 cohorts of 400 men armed in Roman fashion and 2,000 horse.

BIBLIOGRAPHY

F. M. Abel, 'Antiochus Épiphane' in *Vivre et penser, recherches d'exégèse et d'histoire 1 = Revue biblique* 50 (1941) pp. 231-254.

Israel Abrahams, *Campaigns in Palestine from Alexander the Great* (London 1927).

Aelian, Historical Miscellany, Edited and Translated by N. G. Wilson (Loeb ed. Cambridge Mass., London 1997).

Aeneas Tacticus, Asclepiodotus, Onasander, with an English translation by members of the Illinois Greek Club (Loeb ed., Cambridge Mass. 1923)

Antike Gemmen in Deutschen Sammlungen. Bd. I Staatliche Münzsammlung München, Teil 1, ed. Elfreide Brandt (Munich 1970).

Α.Σ. Αρβανιτοπουλου, *Γραπται Στηλαι Δημητριαδος-Παγασων* (Athens 1928).

Asclépiodote Traité de Tactique. Texte établi et traduit par Lucien Poznanski (ed. Budé, Paris 1992).

Alan E. Astin, *Cato the Censor* (Oxford 1978).

M.M. Austin, *The Hellenistic world from Alexander to the Roman Conquest. A selection of ancient sources in translation* (Cambridge etc. 1981).

M.M. Austin, 'Hellenistic Kings, War, and the Economy' *CQ* 36 (1986) pp. 450-66.

Ernest Babelon, *Catalogue des Monnaies grecques de la Bibliothèque nationale, Les Rois de Syrie, d'Arménie et de Commagène* (Paris 1890).

I. Bake, *Posidonii Rhodii Reliquae Doctrinae* (Leiden 1810).

H. R. Baldus, 'Der Helm des Tryphon und die seleukidische Chronologie der Jahre 146-138 v. Chr.' *Jahrbuch für Numismatik und Geldgeschichte* 20 (1970) pp. 217-239 pl. 16.

John R.Bartlett, *The Cambridge Bible Commentary, New English Bible. The First and Second Books of the Maccabees* (Cambridge 1973).

H. I. Bell, 'Notes on Early Ptolemaic Papyri. 1. An Early Ptolemaic Contract of Loan' in ed. Ulrich Wilcken, *Archiv Für Papyrusforschung und Verwandte Gebeite* 7 (Berlin 1924) pp. 17-27.

Bezalel Bar-Kochva, *The Seleucid Army. Organization and Tactics in the Great Campaigns* (Cambridge 1976).

Bezalel Bar-Kochva, *Judas Maccabaeus. The Jewish Struggle against the Seleucids* (Cambridge 1989).

Hermann Bengtson, *Die Strategie in der Hellenistischen Zeit 2* (Munich 1944).

Étienne Bernand, *Inscriptions métriques de l'Égypte gréco-romaine* (Paris 1969).

Étienne Bernand, *Inscriptions grecques d'Hermoupolis Magna et de se nécropole* (Cairo 1999).

Edwyn Robert Bevan, *The House of Seleucus* (London 1902).

Edwyn Bevan, *A History of Egypt under the Ptolemaic Dynasty* (London 1927).

E. Bikerman, *Institutions des Séleucides = Haut-Commissariat de la République Française en Syrie et au Liban, Service des Antiquités, Bibliothèque archéologique et historique Tome XXVI* (Paris 1938).

ed. M. C. Bishop, *The Production and Distribution of Roman Military Equipment* (= *British Archaeological Reports, International Series 275*, Oxford 1985).

Yann Le Bohec, *The Imperial Roman Army* (Oxford 1994)

A.B. Bosworth, 'Arrian's Literary Development' *CQ* n.s. 22 (1972) 163-185.

A.B. Bosworth, *Conquest and Empire. The Reign of Alexander the Great* (Cambridge etc. 1988).

A.B. Bosworth, *From Arrian to Alexander. Studies in Historical Interpretation* (Oxford 1988).

A. Bouché-Leclercq, *Histoire des Lagides IV* (Paris 1907).

E. S. Bouchier, *A Short History of Antioch 300 B.C. - A.D. 1268* (Oxford 1921).

Evaristo Breccia, *Catalogue Général des Antiquités égyptiennes du Musée d'Alexandrie - Iscrizioni Greche e Latine . Vol 1 Nos 1-568* (Cairo 1911).

David J. Breeze, 'The Career Structure below the Centurionate during the Principate' in *Aufstieg und Niedergang der Römischen Welt II. 1* (Berlin 1974) pp. 435-51.

John Briscoe, *A Commentary on Livy Books XXXI-XXXIII* (Oxford 1973).

Kai Broderson, *Appians Antioche (Syriake 1, 1 - 44, 232) Text und Kommentar nebst ein Anhang: Plethons Syriake - Exzerpt etc.* (Munich 1991).

Blanche R. Brown, *Ptolemaic Paintings and Mosaics and the Alexandrian Style* (Cambridge Mass. 1957).

P.A. Brunt, *Italian Manpower 225 B.C. - A.D. 14* (Oxford 1971).

J.G. Bunge, 'Die Feiern Antiochos' IV. Epiphanes in Daphne im Herbst 166 v. Chr.' *Chiron* 6 (1976) pp. 53-71.

Stanley Mayer Burstein, *The Hellenistic Age from the battle of Ipsos to the death of Kleopatra* = edd. E. Badian and Robert K. Sherk, *Translated Documents of Greece and Rome, Volume 3* (Cambridge 1985).

P. Callaghan, 'The Medusa Rondanini and Antiochus III' *BSA* 76 (1981) pp. 59-70.

P.G. Calligas, 'Roman Helmets in Greece' *AAA* 18 (1985) pp. 161-4.

John Camp, Michael Ierardi, Jeremy McInerney, Kathryn Morgan and Gretchen

Umholz, 'A Trophy from the battle of Chaironeia of 86 BC' *AJA* 96 (1992) pp. 443-455.

Brian Campbell, 'Teach Yourself How to be a General' *JRS* 77 (1987) 13-29.

ed. Sheila D. Campbell, *The Malcove Collection. A Catalogue of the Objects in the Lillian Malcove Collection of the University of Toronto* (Toronto Buffalo London 1985).

Constantin Carapanos, *Dodone et ses Ruines* (Paris 1878).

Paul Cartledge & Antony Spawforth, *Hellenistic and Roman Sparta. A Tale of Two Cities* (London, New York 1989).

Fritz Caspari, 'Studien zu dem Kallixeinos Fragment Athenaios 5, 197c - 203b' *Hermes* 68 (1933) pp. 400-414.

D. Cohen, 'Annotationes ad Auctores et Papyros Nonnullas II οἱ ἔξω τάξεων' *Mnemosyne* 54 (1926) pp. 82-3.

Gaston Colin, *Fouilles de Delphes Tome III Épigraphie, Fascicule IV Inscriptions de la Terrasse du Temple et de la Région nord du Sanctuaire Nos. 1-86* (Paris 1930).

Paul Couissin, *Les Institutions militaires et navales* (Paris 1932).

J.J. Coulton, 'Termessians at Oinoanda' *AS* 32 (1982) pp. 115-131.

D.J. Crawford, *Kerkeosiris. An Egyptian Village in the Ptolemaic Period* (Cambridge 1971).

E. Van't Dack, 'Sur l'Évolution des Institutions Militaires Lagides' *Armées et fiscalité dans le Monde antique, Colloques Nationaux du C.N.R.S No, 936* (Paris 1977).

E. Van't Dack, W.Clarysse, G.Cohen, J. Quaegebeur & J.K. Winnicki, *The Judean-Syrian-Egyptian Conflict of 103-101 B.C.. A Multilingual Dossier Concerning a „War of Sceptres"* (*Collectanea Hellenistica I,* Louvain 1989).

E. Van't Dack, *Ptolemaica Selecta. Études sur l'armée et l'administration lagides* (*Studia Hellenistica 29,* Louvain 1988).

Die Daker. Katalog der Ausstellungs im Römisch-Germanisch Museum in Köln (Cologne 1980).

R.W. Daniel, 'The Military σημεῖον in P.Amh. II 39' *ZPE* 52 (1983) pp. 269-71.

Norman Davis and Colin M. Kraay, *The Hellenistic Kingdoms. Portrait Coins and History* (London 1973).

James G. DeVoto (ed., tr.). *Flavius Arrianus ΤΈΞΝΗ ΤΑΚΤΙΚΆ (Technical Handbook) and ''ΈΚΤΑΞΙΣ ΚΑΤÀ 'ΑΛΆΝΩΝ (The Expedition Against the Alans)* (Chicago 1993).

Petros Dintsis, *Hellenistische Helme* (Rome 1986).

Alfred v. Domeszewski, *Die Phalangen Alexanders und Caesars Legionen (= Sitzungsberichte der Heidelberger Akademie der Wissenschaften, Philosophisch-historische Klasse 1925/6. 1*, Heldelberg 1926).

G. A. Downey, *A History of Antioch in Syria from Seleucus to the Arab*

Conquest (Princeton 1961).

Arthur M. Eckstein, *Moral Vision in The Histories of Polybius* (Berkeley 1995).

Henri Van Effenterre, *La Crète et la monde grec de Platon a Polybe* (*Bibliothéque des Écoles françaises d'Athénes et de Rome Fasc. 163,* Paris 1948).

R. M. Errington, *Philopoemen* (Oxford 1969).

Michel Feyel, 'Le Règlement militaire trouvé a Amphipolis' *Rev. arch.* 1935, 2 pp. 29-68.

Mette Fjeldhagen, *Graeco-Roman Terracottas from Egypt* (Ny Carlsberg Glyptotek, Copenhagen 1995).

Victoria Foertmeyer, 'The Dating of the Pompe of Ptolemy II Philadelphus' *Historia* 37 (1988) pp. 90-104.

Emilio Gabba, *Republican Roma, The Army and the Allies* (Oxford 1976).

Ch. Clermont-Ganneau, 'Stèles peintes de Sidon' *Gazette archéologique* 3 (1877) pp. 102-115 pls. 15-16.

Volkmar Von Graeve, *Der Alexander Sarkophag und seine Werkstatt* (=*Istanbuler Forschungen 28,* Berlin 1970).

John D. Grainger, *The Cities of Seleukid Syria* (Oxford 1990).

P. Gardner, *A Catalogue of Greek Coins in the British Museum, Seleucid kings* (London 1878).

M. J. Geller, 'New Information on Antiochus IV from Babylonia Astronomical Diaries' *Bulletin of the School of Oriental and African Studies* 54 (1991) 1-4.

Peter Green, *Alexander to Actium. The Hellenistic Age* (London 1990).

Bernard P. Grenfell, Arthur S. Hunt, David G. Hogarth, *Fayûm towns & their Papyri* (London 1900).

G. T. Griffith, *The Mercenaries of the Hellenistic World* (Cambridge 1935).

E.S. Gruen, *The Hellenistic World and the Coming of Rome* (Berkeley and Los Angeles 1984).

N. G. L. Hammond and G. T. Griffith, *A History of Macedonia. Volume II 550-336 B.C.* (Oxford 1979).

N.G.L. Hammond, *Alexander the Great. King, Commander and Statesman* [2] (Bristol 1989).

G.M.A. Hanfman, 'New Fragments of Alexandrian Wall painting' in *Alessandria e il mondo ellenistico-romano, studi in onore di Achille Adriani 2, = Studi e materiali, Istituto di archeologia Universita Palermo* 5 (Rome 1984) pp. 242-255.

E.M. Hanson, 'Alexander's hypaspists and the argyraspides' *Historia* 30 (1981) 117-20.

ed. Victor Davies Hanson, *Hoplites: The Classical Greek Battle Experience* (1991).

Bruno Helly, *Gonnoi II , Les Inscriptions* (Amsterdam 1973).

B.Helly (ed.), *La Thessalie, Actes de la Table Ronde 21-24 juillet 1974* (Lyon, Paris 1979).

Bruno Helly, *L'État thessalien. Aleuas le Roux les Tétrades et les Tagoi* (Lyon 1995).

Maurice Holleaux, *Études d'Épigraphie et d'Histoire grecques III Lagides et Séleucides* (Paris 1942).

Arthur Houghton, *Coins of the Seleucid Empire from the Collection of Arthur Houghton* (New York 1983).

Ann Hyland, *Training the Roman Cavalry. From Arrian's Ars Tactica* (London 1993).

Louis Jalabert, 'Nouvelles Stèles peintes de Sidon' *Rev. arch.* 1904, 2 pp. 1-16.

C.P.Jones, *Culture and Society in Lucian* (Cambridge Mass. London 1986).

P. Jouget,'Inscriptions grecques d'Égypte III. Hermopolis Magna (*Ashmounein*)' *BCH* 20 (1896) pp. 177-191.

I.G. Kidd, *Posidonius II. The Commentary (i)* (Cambridge 1988).

Franz Kiechle, "Die 'Taktik' des Flavius Arrianus", *Bericht der römisch-germanischen Kommission* 45 (1965) 87-129.

Wolfgang Kimmig,'Ein Keltenschild aus Äegypten' *Germania* (1940) pp. 106-111, pl 23.

Walther Kolbe, 'Die Seleukidenära des I. Makkabäerbuches' *Hermes* 62 (1927) pp. 225-242.

J. Kromayer, G. Veith, *Antike Schlachtfelder in Griechenland* 2 (Munich 1907).

Johannes Kromayer, Georg Veith, *Heerwesen und Kriegführung der Griechen und Römer* (= edd. Iwan von Müller, Walter Otto, *Handbuch der Altertumswissenschaft IV, 3, 2*, Munich 1928).

edd. Amélie Kuhrt & Susan Sherwin-White, *Hellenism in the East* (London 1987).

Marie Laffranque, *Poseidonios d'Apamée. Essai de Mise au Point* (Paris 1964).

F. Lammert, 'Σημεῖα' *PW* ii A, 2 (1923) col. 1331-9.

Marcel Launey, *Recherches sur les Armées hellénistiques* I-II (Paris 1949, 1950).

A.W. Lawrence, *Later Greek Sculpture and its Influence on East and West* (London 1927).

J.F. Lazenby and D. Whitehead, 'The Myth of the Hoplite's Hoplon' *CQ* n.s. 46 (1996) pp. 27-33.

Gustave Lefebvre, 'Inscription Grecque d'Ashmounein' *BSAAlex* 10 (1908) pp. 187-195.

Leon Legrain, *Terra-Cottas from Nippur* (*University of Pennsylvania, The University Museum, Publications of the Babylonian Section Vol. xvi*, Philadelphia 1930).

Marie Thérèse Lenger, *Corpus des Ordonnances des Ptolémées* (Brussels 1964).

Th. Lenschau, 'Nikanor (16a)' *PW Supplbd.* vii (1940) col. 568-9.

J. Lesquier, 'Note sur une inscription d'Ashmounein' *Rev. phil.* 32 (1908) pp. 215-25.

Jean Lesquier, *Les Institutions militaires de l'Égypte sous les Lagides* (Paris 1911).

Pierre Lévèque, *Pyrrhos* (= *Bibliothéque des Écoles françaises d'Athénes et de Rome Fasc. 185*, Paris 1957).

Naphtali Lewis, *Greeks in Ptolemaic Egypt* (Oxford 1986).

R. A. Lock, 'The origins of the Argyraspids' *Historia* 26 (1977) 373-8.

Th. Macridy-Bey, 'A travers les néropoles sidoniennes 1. - Stéles peintes de Sidon' *Revue biblique* 1 (1904) pp. 546-556.

Marianne Maaskant-Kleibrink, *Catalogue of the Engraved Gems in the Royal Coin Cabinet The Hague. The Greek and Roman Collections* (The Hague 1978).

J.P.Mahaffy, *A History of Egypt under the Ptolemaic Dynasty* (London 1898).

Jürgen Malitz, *Die Historien des Poseidonios* (= *Zetemata* 79, Munich 1983).

Hugh J. Mason, *Greek Terms for Roman Institutions* (*American Studies in Papyrology Vol. 13,* Toronto 1974).

Gustave Mendel, *Musées Impériaux Ottomans. Catalogue des sculptures grecques, romaines et byzantines, 1* (Istanbul 1912).

J.G. Milne, *Catalogue général des antiquités égyptiennes du Musée d'Alexandrie. Greek Inscriptions* (Oxford 1905).

Robert D. Milns 'The Army of Alexander the Great' in ed. E. Badian, *Alexandre le Grand. Image et réalité. Entretiens Hardt XXII* (Geneva 1976) pp. 87-136.

T. B. Mitford, 'Seleucus and Theodorus' *Opus. Athen.* 1 (1953) pp. 130-171.

T. B. Mitford, 'The Hellenistic Inscriptions of Old Paphos' *BSA* 56 (1961).

Otto Mørkholm, *Studies in the Coinage of Antiochus IV of Syria* (Historisk-filosofiske Meddelelser udgivet af Det Kogelige Danske Videnskabernes Selskab, Bind 40 nr. 3, Copenhagen 1963).

Otto Mørkholm, *Antiochus IV of Syria* (Copenhagen 1966).

Leon Mooren, *The Aulic Titulature in Ptolemaic Egypt, Introduction and Prosopography* (Brussels 1975).

Luigi Moretti, *Iscrizioni Storiche Ellenistiche* I-II (Florence 1967, 1975).

K.K. Müller, 'Aelianus (10)' *PW* i. 1 (1894) col. 482-6.

K.K. Müller, 'Asklepiodotus (10)' *PW* ii (1896) col. 1637-41.

A. K. Narain, *The Indo-Greeks* (Oxford 1957).

Valeri P. Nikonorov and Sergei A. Savchuk, 'New Data on Ancient Bactrian Body-Armour (in the light of Finds from Kampyr Tepe)' *Iran* 30 (1992) pp. 49-54 pl. 14.

Valerii P. Nikonorov, *The Armies of Bactria 700 BC - 450 AD* (Stockport 1997).

A.D. Nock, 'Notes on Ruler-Cult, I-IV' *JHS* 48 (1928) 21-43 = ed. Z. Stewart, *Arthur Darby Nock. Essays on Religion and the Ancient World etc.* (Oxford 1972) I pp. 134-159.

John Paddock, 'Some changes in the manufacture and supply of Roman helmets under the late Republic and early Empire' in ed. M. C. Bishop, *The Production and Distribution of Roman Military Equipment* (= BAR International Series 275, Oxford 1985) p. 142-59.

Eliezer Paltiel, 'Antiochus IV and Demetrius I of Syria' *Antichthon* 13 (1979) pp. 42-47.

R. Parker and W. Dubberstein, *Babylonian Chronology 626 B.C. - A.D. 75* (= *Brown University Studies 19*, Providence 1956).

Klaus Parlasca, *Syrische Grabreliefs hellenistischer und römischer Zeit. Fundgruppen und Probleme* (= *Trier Winckelmannsprogramme 3,* 1981).

P. Pédech, *La méthode historique de Polybe* (Paris 1964).

Paul Pédech, *Polybe, Histoires I* (ed. Budé, Paris 1969).

Werner Peek, *Griechische Vers-Inschriften I* (Berlin 1955).

W. Peremans & E. Van't Dack, *Prosopographica* (*Studia Hellenistica 9*, Louvain, Leiden 1953).

Poland, 'Σπεῖρα' *PW* iii A, 2 (1929) col. 1586-1592.

Lucien Poznanski, 'Essai de Reconstruction du *Traité de Tactique* de Polybe d'après le Livre III des *Histoires*' *L'Antiquité classique* 49 (1980) pp. 161-172.

W. Kendrick Pritchett, *Studies in Ancient Greek Topography Part II (Battlefields)* (Berkeley and Los Angeles 1969).

Elizabeth Rawson, *Intellectual Life in the Late Roman Republic* (London 1985).

A.J. Reinach, 'Incriptions de Itanos' *RÉG* 24 (1911) pp. 377-425.

Karl Reinhardt, 'Poseidonios (3)' *PW* xxii, 1 (1953) col. 558-826.

Fritz Reuter, *Beiträge zur Beurteilung des Königs Antiochos Epiphanes* (diss. Münster 1938).

E.E. Rice, *The Grand Procession of Ptolemy Philadelphus* (Oxford 1983).

G.M.A. Richter, *The Portraits of the Greeks Volume Three* (London 1965).

G.M.A. Richter, Abridged and revised by R.R.R. Smith, *The Portraits of the Greeks* (Oxford 1984).

Georges Le Rider, *Suse sous les Séleucides et les Parthes* (= *Mémoires de la Mission archéologique en Iran* 38, Paris 1965).

Georges Le Rider, 'Les resources financières de Séleucos IV (187-175) et le paiement de l'indemnité aux Romaines' in edd. Martin Price, Andrew Burnett and

Roger Bland, *Essays in Honour of Robert Carson and Kenneth Jenkins* (London 1993) pp. 49-68.

Louis Robert, *Études anatoliennes* (= *Études orientales publiées par l'Institut Français d' Archéologie de Stamboul 5,* Paris 1937).

Louis Robert, *Hellenica. Recueil d'épigraphie, de numismatique, et d'antiquités grecques* 1-13 (Paris 1940-1965).

L. Robert, *Opera Minora Selecta* 1-4 (Amsterdam 1969-1974).

L. Robert, *Noms indigènes dans l'Asie-Mineure gréco-romaine (Bibliothèque archéologique et historique de l'Institut frrançais d'Archéologie d'Istanbul,* Paris 1963).

H. Russell Robinson, *The Armour of Imperial Rome* (New York 1975).

M. Rostovtzeff, *The Social and Economic History of the Hellenistic World*[2] (Revised edition, Oxford 1953).

P. Roussel, 'La population de Délos a la fin du IIe. siècle av. J. -C.' *BCH* 55 (1931) pp. 438-449.

P. Roussel, 'Un Règlement militaire de l'Époque macédonienne' *Rev. arch.* 1934, 1 pp. 39-47.

Rainer Roy and Friedrid Kohler, 'Festaufzug, Festeinzug' in *Reallexicon zur deutschen Kunst-Geschichte* Vol. 8 (1987) col. 1417-1520.

Alan E. Samuel, *Greek and Roman Chronology. Calendars and Years in Classical Antiquity* (= edd. Iwan von Müller, Walter Otto, Herman Bengtson, *Handbuch der Altertumswissenschaft I, 7;* Munich 1972).

Gaetano De Sanctis, 'Epigraphica XII. Il regolamento militare dei Macedoni' *Riv. Fil.* 12 (1934) pp. 515-21.

Valdemar Schmidt, *De Graesk-Aegyptiske Terracotter I Ny Carlsberg Glyptothek* (Copenhagen 1911).

Paul Schoh, 'Nikanor (16)' *PW* xvii, 1 (1936) col. 270.

Nick Sekunda and Angus McBride, *The Army of Alexander the Great* (London 1984).

Nick Sekunda, Angus McBride, *Republican Roman Army 200-104 BC* (London 1996).

Henri Seyrig, *Notes on Syrian Coins* (= *American Numismatic Society, Numismatic Notes and Monographs* no. 119, New York 1950).

S. M. Sherwin-White, 'A Greek Ostrakon from Babylon of the Early Third Century BC' *ZPE* 47 (1982) pp. 51-70.

Sherwin-White & Kuhrt, *Samarkand to Sardis. A New Approach to the Seleucid Empire* (London 1993).

J. Gilbart Smyly, *Greek Papyri from Gurob* (1921).

Michael Speidel, *The Framework of an Imperial Legion* (The Fifth Annual Caerleon Lecture, Cardiff 1992).

Augustin Spendel, *Untersuchungen zum Heerwesen der Diadochen* (Breslau/Wroclaw 1915).

Alk. Spetsieris, 'Early Roman Bronze Helmet from Polydrosso in Thresprotia' *AAA* 15. 1 (1982) pp. 59-64.

Stylianos Spyridakis, *Ptolemaic Itanos and Hellenistic Crete* (= *University of California Publications in History, Volume 82*, Berkeley Los Angeles London 1970).

Philip A. Stadter, 'The *Ars Tactica* of Arrian: Tradition and Originality' *CP* 73 (1978) pp. 117-128.

Philip A. Stadter, *Arrian of Nicomedia* (Chapel Hill 1980).

Stähelin, 'Lykortas' *PW* xiii, 2 (1927) col. 2386-9.

Friedrich Stählin, *Das hellenische Thessalien* (Stuttgart 1924).

Dr. Modestus Van Straaten O.E.S.A., *Panétius, sa vie, ses écrits et sa doctrine avec une édition des fragments* (Amsterdam 1946).

Roy Strong, *Art and Power, Renaissance Festivals 1450-1650* (Woodbridge Suffolk 1984).

F. Susemihl, *Geschichte der griechischen Literatur in der Alexandrinerzeit* I-II (Leipzig 1891-2).

Mary Hamilton Swindler, *Ancient Painting from the Earliest Times to the Period of Christian Art* (New Haven 1929).

W.W. Tarn, *Alexander the Great II. Sources and Studies* (Cambridge 1948).

W. W. Tarn, 'Asclepiodotus'.*OCD*[2] (London 1970) p. 129.

W.W. Tarn, *The Greeks in Bactria and India*[2] (Cambridge 1951).

Mario Toarelli, *Typology and Structure of Roman Historical Reliefs* (1982).

H. Tonnet, *Recherches sur Arrien. Sa Personalite et ses ecrits atticistes* (Amsterdam 1988).

Augustus Traversa, *Index Stoicorum Herculanensis* (Rome 1952).

Fritz Uebel, *Die kleruchen Ägyptens unter den ersten sechs Ptolemäern* (Berlin 1968).

Eugene Vanderpool, 'The Marble Trophy from Marathon in the British Museum' *Hesp.* 36 (1967) pp. 108-111, pl. 31.

Volkman, 'Ptolemaios (24)' *PW* 23, 2 (1959) col. 1702-3.

F.W. Walbank, *Philip V of Macedon* (Cambridge 1940).

F.W. Walbank, *A Historical Commentary on Polybius* I (Oxford 1956), II (Oxford 1967), III (Oxford 1979).

F.W. Walbank, *Polybius* (*Sather Classical Lectures Volume 42*) (Berkeley, Los Angeles, London 1972).

F.W. Walbank, *The Hellenistic World* (London 1981).

F.W. Walbank, 'Supernatural Paraphernalia in Polybius' *Histories*' in ed. Ian Worthington, *Ventures into Greek History* (Oxford 1994) pp. 28-42.

F.W. Walbank, 'Two Hellenistic Processions: A Matter of Self-Definition' in edd. Hannah M. Cotton, Jonathan J. Price, David Wasserstein *Studies in Memory of Abraham Wasserstein Volume 1* = *Scripta Classica Israelica* 15 (1996) pp. 119-130.

H.B. Walters, *Catalogue of the Terracottas in the Department of Greek and Roman Antiquities, British Museum* (London 1903).

G. R. Watson, *The Roman Soldier* (London 1969).

Graham Webster, *The Roman Imperial Army of the First and Second Centuries A.D.* (second edition, London 1979).

C. Bradford Welles, *Royal Correspondence in the Hellenistic Period. A Study in Greek Epigraphy* (London 1934).

M. Wellmann, 'Aelianus (10)' *PW* i, 1 (1894) col. 486-8.

H. M. Werner, *De Polybii vita et itineribus quaestiones chronologicae* (Diss. Leipzig 1877) pp. 15-16 (non vidi).

Everett L. Wheeler, 'The Occasion of Arrian's *Tactica*' *GRBS* 19 (1978) 351-365.

Everett L. Wheeler, 'The Legion as Phalanx' *Chiron* 9 (1979) pp. 303-318.

Everett L. Wheeler, *Stratagem and the Vocabulary of Military Trickery* (*Mnemosyne Suppl. 108*, Leiden 1988).

Edouard Will, *Histoire politique du Monde hellénistique (323 - 30 av, JC)* 2 (Nancy 1982).

Rolf Winkes, 'The Pyrrhus Portrait' in edd. Tony Hackens, Nancy D. Holloway, R. Ross Holloway and Ghislane Moucharte, *The Age of Pyrrhus* (*Archaeologia Transatlantica* XI, Louvain-La-Neuve, 1992).
Everett L. Wheeler, 'The Legion as Phalanx' *Chiron* 9 (1979) pp. 303-318.
A. Wilhelm, Αἰγυπτιακά (*Sitz. Akad. Wien,* Vienna 1946).

Ulrich Wilken, *Griechische Ostraka aus Aegypten und Nubien* II (Leipzig, Berlin 1899).

Ulrich Wilcken, *Archiv für Papyrusforschung* 1- (Leipzig 1901-).

Jan Krzysztof Winnicki, *Ptolemäerarmee in Thebais* (Wroclaw 1978).

Jan Krzysztof Winnicki, 'Ein Ptolemäischer Offizer in Thebais' *Eos* 60 (1972) pp. 343-53.

M. Wörrle, *Stadt und Fest im kaiserzeitlichen Kleinasien* (*Vestigia vol. 39*, 1988).

Yigael Yadin, *The Scroll of the War of the Sons of Light against the Sons of Darkness* (Oxford 1962).

Marcello Zambelli, 'L'Ascesa al Trono di Antioco IV' *Riv. Fil.* 88 (1960) pp. 363-389.

F. Ziegler, 'Polybios (1)' *PW* xxi, 2 (1952) col. 1440-1578.

Ladislav Zgusta, *Kleinasiatische Personnennamen* (Prague 1964).

F. Zucker, *Doppelinschrift spätptolemâischer Zeit aus der Garnison von Hermopolis Magna* (= *Abhandlungen der Preußischen Akademie der Wissenschaften, Jahrgang1937, Philosophisch-historische Klasse Nr. 6*, Berlin 1938).

Friedrich Zucker, 'Nachträge zur 'Doppelinscrift spätptolemäischer Zeit aus der Garnison von Hermopolis Magna' *Aegyptus* 18 (1938) pp. 279-284.